W9-DFH-992

DEMCO, INC. 38-2931

The Hindbrain

GRAY
MATTER

GRAY
MATTER

The Hindbrain

Carl Y. Saab

Assistant Professor—Research

Department of Surgery

Brown University

CHELSEA HOUSE
PUBLISHERS

A Haights Cross Communications ✈ Company ®

Philadelphia

CHELSEA HOUSE PUBLISHERS

VP, NEW PRODUCT DEVELOPMENT Sally Cheney
DIRECTOR OF PRODUCTION Kim Shinners
CREATIVE MANAGER Takeshi Takahashi
MANUFACTURING MANAGER Diann Grasse
PRODUCTION EDITOR Noelle Nardone
PHOTO EDITOR Sarah Bloom

STAFF FOR THE HINDBRAIN

PROJECT MANAGEMENT Dovetail Content Solutions
DEVELOPMENTAL EDITOR Carol Field
PROJECT MANAGER Pat Mrozek
PHOTO EDITOR Deanna Laing
SERIES AND COVER DESIGNER Terry Mallon
LAYOUT Maryland Composition Company, Inc.

A Haights Cross Communications ✦ Company ®

www.chelseahouse.com

First Printing

10 9 8 7 6 5 4 3 2 1

Library of Congress Cataloging-in-Publication Data

Saab, Carl Y.
    Hind brain / Carl Y. Saab.
    p. cm. — (Gray matter)
Includes bibliographical references.
    ISBN 0-7910-8510-4
1. Rhombencephalon. I. Title. II. Series.
QP376.S22 2005
612.8′27—dc22   2005011687

Dedication
To all animals sacrificed for biomedical research.

Contents

1 | The Hindbrain: The Mediator

If we consider the body to be a group of integrated systems of tissues (e.g., the blood, the skin, the muscle, and the **neurons**), the brain would be in a good position to control and coordinate the function of all of these systems. Within this context, if we think of the brain as the "chief commander," we could think of the spinal cord as the soldier that carries out the voluntary commands of the brain. The hindbrain is a mediator between the commander and the soldier. In fact, the hindbrain not only relays neuronal signals between the brain and the spinal cord but also serves as a modulator of some of the voluntary commands before their execution.

INTRODUCTION TO HINDBRAIN FUNCTIONS

It is difficult to imagine what the life of a human being would be like without the hindbrain. The hindbrain is not a simple link between the brain and the spinal cord, although anatomically it is interposed between these two structures. Many neurons located within the hindbrain play a critical role in regulating vital functions such as breathing, heartbeat, swallowing, eye movements, and coordination of motor behavior ranging from swinging a baseball bat to maintaining balance and stabilizing gaze during head rotations. In addition, a major component of the hindbrain, the cerebellum, is not interposed be-

1

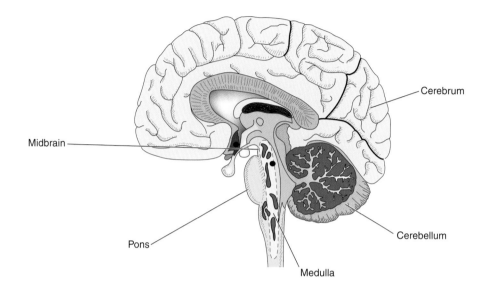

Figure 1.1 This midsagittal view of the brain highlights the hindbrain: the cerebellum, pons, and medulla.

tween the spinal cord and the brain, but rather is located in the back part of the base of the brain.

WHAT AND WHERE IS THE HINDBRAIN?

The hindbrain includes the cerebellum, the pons, and the medulla (Figure 1.1). These structures belong to the **central nervous system** and function together and with the rest of the nervous system to support and coordinate vital bodily functions (Figure 1.2). The medulla is the lowest part of the hindbrain and joins the hindbrain to the spinal cord. The medulla controls unconscious, yet essential, body functions such as respiration, swallowing, and muscle tone. The pons is located above the medulla. The pons acts as a bridge between neurons in the **nuclei** of the **brain stem** and the cerebellum and receives neuronal sig-

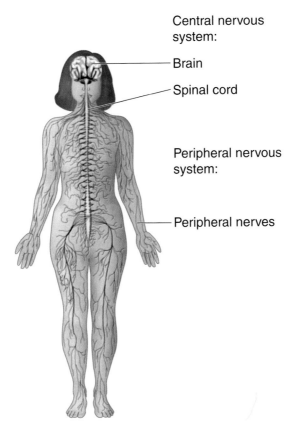

Central nervous system:

Brain

Spinal cord

Peripheral nervous system:

Peripheral nerves

Figure 1.2 The central nervous system consists of the brain and spinal cord, whereas the peripheral nervous system refers to neuronal tissue outside the skull and the vertebral column (e.g., peripheral nerves).

nals descending from other brain areas. The pons carries information that is important for the control of eye movements. Along with the cerebellum, the pons helps coordinate voluntary eye and body movements. Above and behind the pons is the cerebellum, our "little brain." The cerebellum is by far the largest structure of the hindbrain and contains the most neurons. In fact, the cerebellum contains approximately half the neurons in the entire central nervous system. Although the cerebellum does not underlie vital physiological functions, it does modulate fine body movements that play a role in complex human behaviors and promote human evolution. For example, the cerebellum supports fine hand movements (for making tools for hunting and gathering), communication by language (speech articulation), and other arguably distinctive human behavior (e.g., acrobatics and the technical aspects of playing musical instruments involving hand and finger coordination and accurate timing).

■ **Learn more about neural control of movement** Search the Internet for *motor coordination*, *cerebellum*, and *basal ganglia*.

2 Cellular Components and General Organization of the Nervous System

COMPASS TO NERVOUS SYSTEM ANATOMY

North, south, east, and west are great compass points to know when navigating a map. They are useless, though, when navigating the nervous system. Special terms allow us to describe direction within the nervous system (Figure 2.1). These terms, mostly Latin in origin, include rostral (*rostralis*), toward the brain; caudal (*caudalis,* or "tail"), away from the brain; dorsal (*dorsalis,* or "back"), toward the back of the body; and ventral (*ventralis,* or "belly"), toward the front of the body. In addition, nervous system tissue can be cut across different planes to study its cellular organization. These cuts can be horizontal (parallel to the ground), coronal (referring to how a crown is placed on the head, or "sideways"), and sagittal (perpendicular to the coronal plane in a front-to-back orientation).

The nervous system consists of cells that are some of the longest in the body: neurons (nerve cells) (Figure 2.2). A group of neurons densely gathered within a clearly delineated space in the nervous system tissue and sharing comparable morphology (shape as seen under a microscope) is referred to as a nucleus. Neurons relay their information in the form of electrochemical activity to other neurons by forming **synapses**. Neurons also require other cells called **glia** for structural and nutritional support. Neurons have a cell body

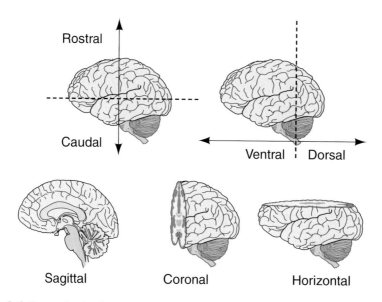

Figure 2.1 To navigate the nervous system, you need to understand the terms that describe orientation of the body: rostral (*rostralis*), meaning toward the brain; caudal (*caudalis,* or "tail"), meaning away from the brain; dorsal (*dorsalis,* or "back"), meaning toward the back of the body; and ventral (*ventralis,* or "belly"), meaning toward the front of the body. Orientation is also described by the terms *horizontal,* meaning parallel to the ground; *coronal,* meaning how a crown is placed on the head, or "sideways"; and *sagittal,* meaning perpendicular to the coronal plane in a front-to-back orientation.

containing the machinery for making proteins necessary for the neuron's survival and function. The cell body makes molecules called neurotransmitters. **Neurotransmitters** are then transported to the terminal endings of a neuron where they may be stored in vesicles and released in the synapse to impact other neurons or other target organs (e.g., a muscle or a gland).

An extension or "axon" projects from the cell body. Some axons are wrapped in a fatty sheath called **myelin**. This sheath insulates the axon for physical protection and faster electrical conduction. Dense myelin appears white (because it's fatty), whereas neuronal cell bodies appear gray when a brain specimen is examined.

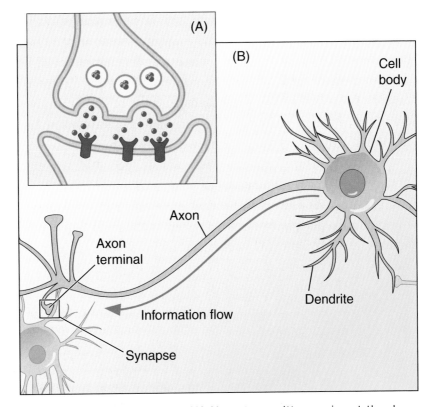

Figure 2.2 A typical neuron. (A) Neurotransmitters arrive at the dendrites, where they bind to receptors and cause tiny electrical currents. (B) If they are strong enough, these currents generate action potentials that travel down the axon toward the terminal branch. The myelin sheath, composed of Schwann cell processes in the peripheral nervous system (or oligodendrocytes in the central nervous system), protects and insulates the axon and helps the electrical impulses travel faster.

Accordingly, the term *white matter* refers to bundles of axons (e.g., the outside appearance of the spinal cord), whereas *gray matter* refers to cell bodies (e.g., a nucleus inside the brain or the spinal cord). If the axon terminal of a neuron is disconnected from its cell body (e.g., after **spinal cord injury** or amputation of a limb), the distal part of the neuron (the part of its axon that's disconnected from its injured cell body) will degenerate. This is the process of Wallerian degeneration first described by Augustus Waller in 1852 (see "Augustus Volney Waller" box). Neuronal degeneration will be discussed again later within the context of linking motor or sensory deficits to neurological damage. If the cell body survives the injury, it may regenerate in the peripheral nervous system but not in the central nervous system (with rare exceptions).

Communication between neurons is possible through synapses. Following injury, axons may be damaged and, thus, synapses degrade and communication between neurons is interrupted. In the peripheral nervous system, axons may regenerate, forming new synapses. However, axons in the central nervous system are very resistant to regenerative attempts and, therefore, spinal cord or brain injuries cause irreversible damage. Much of

Augustus Volney Waller

Augustus Volney Waller, the son of William Waller, was born in England in 1816, but lived in the South of France until his father's death in 1830. At age 14, he returned to school in England to study the physical sciences. Waller later studied medicine in Paris, receiving his M.D. degree in 1840. He first described the process of Wallerian degeneration, which was based on his studies of the nerves of a frog's tongue, in 1846. When Waller cut these nerves, he found that the distal stump of the neuron would degenerate.

Augustus Volney Waller (1816–1870)

the research focused on cures for central nervous system injury deals with ways to enhance **axonal regeneration**.

■ **Learn more about axons** Search the Internet for *myelin sheath* or *neuronal axon*.

THE CENTRAL NERVOUS SYSTEM

The central nervous system consists of the spinal cord and the brain, whereas the peripheral nervous system refers to all other neurons, mainly located outside of the skull and the vertebral column. The spinal cord connects the brain with the rest of the nervous system. Therefore, the central nervous system may be classified into different compartments, but it is nevertheless a continuous system with the rest of the peripheral nervous system. Accordingly, many neurons in the brain project to the lower body via axons traveling in the spinal cord, and many neurons in the spinal cord project to the brain via axons traveling in the spinal cord. There is continuity throughout the nervous system, and no distinct demarcation line can be drawn between the spinal cord and the brain to separate them into two independent compartments or functional entities without causing Wallerian degeneration. Similarly, the spinal cord is not completely independent from the peripheral nervous system for comparable reasons; for example, damage to the radial nerve that supplies the hand region may affect neurons in the upper spinal cord. For academic simplification, however, the central nervous system refers to the brain and spinal cord, with the rostral "virtual" demarcation line between the spinal cord and the brain in the lower brain stem at the level of the lower medulla.

ELECTROPHYSIOLOGY

One basic principle of life is change, and neurons help living organisms detect and respond to these changes quickly for adaptation and survival. Although neurons are not the only cells that can respond to environmental cues, they are one of the few types of living cells capable of communicating by electric current, and are hence classified as "excitable" cells, able to produce electricity. Electric current produced by neurons can be measured; in fact, it can also be a reliable indicator of their state of activity. How can we catch neurons in action?

It is very rare that a physical event is so directly linked to a biological or physiological phenomenon, as in the case of neurons. Here we face a living system, a biological tissue capable of producing physical energy in the form of current, which, if we could directly measure it, can give us valuable insight into its function. Because neurons make up the brain and the spinal cord, this measure, called **electrophysiology**, may reveal to us the workings of the brain, perhaps even the mind. Although regarded as highly sophisticated, electrophysiology relies on simple basic techniques. Neurons cannot be seen with the naked eye without a microscope, so it is logical to assume that the miniature current they produce cannot be recorded by conventional tools such as an **ammeter**. A sensor needs to be brought close to a neuron for extracellular recording (of course, this must be done carefully to minimize tissue damage, especially in the brain) or inside a neuron for intracellular recording. That sensor needs to be connected to an amplifier to boost the neuronal signal. The amplifier is generally in the form of a thin needle, referred to as an electrode. If the sensor is very thin, it is referred to as a microelectrode, with perhaps the diameter of a hair. Typical electrodes are made of metal or glass; for metal electrodes, copper or tungsten wires are used. Glass electrodes, or capillaries (thin cylinders 1 to 2 millimeters in diameter), are stretched under high temperature to yield thin microelectrodes of several hundredths of a micrometer in diameter (one micrometer is one thousandth of a millimeter). Since metal wire electrodes conduct electric current readily (low resistance) and those made of glass are non-conductive (very high resistance), glass electrodes can be filled with a solution (conductive), and a metal wire then inserted inside the glass capillary. Both types of electrodes are then connected to an amplifier to boost the neuronal signal (electricity in the form of current). Often, that same signal is also connected to an audio system that allows researchers to "hear" neuronal activity, while that same activity can simultaneously be "seen" on an **oscilloscope** (which works on the same principle as a TV screen). Obviously, monitoring brain activity is not trivial and usually requires recording such activity on a computer for thorough analysis.

OTHER METHODS OF RECORDING NEURONAL ACTIVITY

Electrophysiology is a "real-time" indicator of neuronal activity using physical techniques (measurement of electric current, or in special cases, voltage changes). Other methods that rely on physical techniques that could also be used are based on electromagnetic principles, in particular **functional magnetic resonance imaging** (fMRI). Using fMRI allows researchers first to scan inside the brain or the spinal cord and to localize areas of increased activity; this is based on sensors that, unlike electrodes, are noninvasive (do not need to be inserted into the tissue) and detect minimal changes in oxygen or glucose consumption. Because oxygen and glucose are continuously being used by neurons (in fact, deprivation of glucose and oxygen for more than three minutes may result in neuronal death), the scans obtained by fMRI thus reveal areas of high (or low) neuronal consumption of oxygen or glucose, which indicates high (or low) activity. However, fMRI scans provide only a rough estimate of neuronal activity within a certain region, rather than within single neurons (as in electrophysiology). In addition, fMRI scans, although they provide marvelous anatomical pictures that are

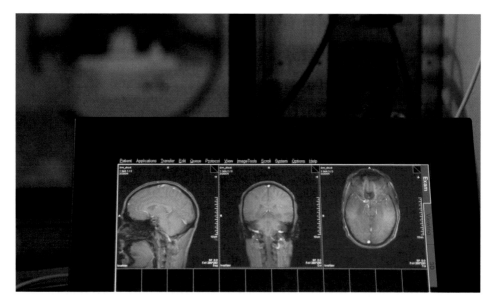

Figure 2.3 The display screen during the scanning process of functional magnetic resonance imaging is shown from the view of the control room. Functional magnetic resonance imaging is used to visualize brain function by showing changes in chemical composition of brain areas or changes in the flow of fluids that occur over time spans of seconds to minutes.

correlated with neuronal activity without surgery, take time to process and show only individual time points of brain activity, rather than a continuous recording (similar to "taking pictures" of an event by fMRI instead of continuously filming it by electrophysiological recording) (Figure 2.3).

EXAMPLES OF NEURONAL RECORDING OR STIMULATION

The brain and the spinal cord are well protected by bone (the skull for the brain and the vertebral column for the spinal cord), and getting access to neurons in the central nervous system requires invasive techniques (under surgery). Therefore, recording neuronal activity is not common in humans for ethical considerations; the benefits do not outweigh the risks. However, in rare cases, neurosurgeons may rely on electrophysiology methods to verify fine anatomical landmarks in the brain during a sensitive brain surgery. For example, one treatment for **Parkinson's disease** requires stimulation of specific brain nuclei (basal ganglia) to compensate for the deficiency in a neurotransmitter (dopamine). In addition to the imaging for accurate localization of the electrode, minimal current pulses are also delivered through the electrode to stimulate neurons within the vicinity of the electrode tip. Depending on the behavioral reaction of the patient and the exact location of the electrode, the anatomy and the function of the desired area to be treated can then be verified. Another therapeutic clinical method relying on electrophysiology is chronic stimulation of nerves (bundle of axons) for the relief of pain, which the patient can "turn on" at will.

3 The Cerebellum: The Little Brain Maestro

Cerebrum means "brain," and *cerebral* refers to "brain matters" or intellect, whereas *cerebellum* means the "little brain." The human brain contains both a cerebrum and a cerebellum. Does this mean we have a big brain and a small brain in our heads? Although these terms may be misleading, they are used mainly to distinguish between two structures in the brain that are intricately connected but look different and occupy different spaces in the skull. The cerebellum, our little brain, occupies the dorso-caudal base of the brain; it is an integral part of the whole brain and, in fact, it coordinates many of the brain's functions, mainly motor commands initiated in the cerebral motor cortex, such as walking or speech articulation. Although it is hard to believe that a section taken from the cerebellum of a fish cannot be easily distinguished under a microscope from that of a human, the cerebellum is in fact so similar across a wide variety of species that a **postmortem** cut through the cerebellum in any normal animal will look alike under a microscope.

THE CEREBELLUM IS NOT SO LITTLE

Although the cerebellum makes up about 25% of the brain, neurons of the cerebellum comprise approximately half of all the neurons in the central nervous system (i.e., the brain and the spinal cord combined), yet they are contained in a relatively small volume. Therefore, the cerebellum is a very compact structure. Its densely packed cells and highly folded surface account for its small volume yet very large surface area (the high surface area-to-volume ratio principle or, in rough terms, more efficient use of space). Throughout perhaps hundreds of millions of years, the cerebellum in more highly evolved species became more compact relative to the rapidly developing cerebrum. The cerebrum, especially the cortex (gray matter of neuronal cell bodies), grew in size (presumably to account for the evolution in primate intelligence, since the cerebral cortex in humans underlies intellect). The cerebellum accommodated this expansion by increasing neuronal density at the expense of volume.

■ **Learn more about the evolution of the cerebellum** Search the Internet for *brain evolution* or *comparative neuroanatomy.*

The cerebellum first evolved over 450 million years ago, well before the emergence of the first vertebrates, and is thus a characteristic part of the brain in cartilaginous fish and pre-vertebrates, as well as in fish, amphibians, reptiles, birds, and mammals. Specifically, the cerebellum evolved out of a specific group of brain stem nuclei. The cerebellum, however, began as a rudimentary appendage to the brain stem, as is evident in lampreys, and then progressively increased in size in the progression from fish to mammals. The cerebellum at first received information regarding balance and a primitive form of hearing via vibration. Then, as animals took to living on dry land, it began to progressively expand to process more complex visual, auditory, and other bodily sensory stimuli. Moreover, the cerebellum began to expand in parallel with the increased size and complexity of the rest of the brain.

Therefore, a primary concern of the cerebellum has been and continues to be stabilizing the body and providing information about the position and movement of the head in relation to gravity. Initially, however, during the early stages of evolution, since primitive creatures had no limbs, they possessed only a small version of the cerebellum (an even smaller "little brain"). This tissue presumably acted to coordinate the muscles that stabilized the head, trunk, and eyes, relative to body position. With the evolution of legs, the cerebellum was forced to take on new roles, including the coordination of other **skeletal muscles**, such as those of the limbs. It is as though the cerebellum relied on the physical principle of "compactness" and resorted to folding its surface within the lower rear part of the skull (posterior cranial fossa), whereas the cerebrum "pushed" the frontal and parietal bones of the skull forward and outward and expanded in volume as well as in area (frontal expansion generally correlates with higher intellect among different species). As a result, the cerebrum in humans looks bigger in size but less folded compared with the cerebellum, which is smaller in size yet much more folded. From these simple anatomical observations emerged a saying relating structure to function. Someone who is not very intelligent is said to have "brains as smooth as a beach pebble," which probably relates to fewer folds in the brain, and presumably fewer neurons and inferior intellectual abilities. However, with regard to this specific example, bear in mind that intelligence is certainly not a function of brain size or neuronal density alone. It is the functional organization of synapses and the efficient use of the available neurons that reinforces intelligence, rather than just size and density.

CEREBRAL AND CEREBELLAR TERRAINS

Foldings or troughs in the cerebral surface are referred to as sulci (singular is **sulcus**), whereas outgrowths or protrusions are known as gyri (singular is **gyrus**) (Figure 3.1). Sulci and gyri in the cerebrum are comparable among people of similar ages. Accordingly, gyri are used as landmarks on the vast and complicated sur-

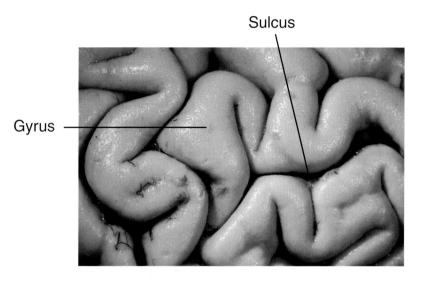

Figure 3.1 This photograph shows the surface of the brain, highlighting the sulci (foldings or troughs) and gyri (outgrowths or protrusions).

face of the cerebrum. For example, the central sulcus marks the division between the primary motor and primary **somatosensory** areas of the cerebral cortex. This area occupies roughly a comparable zone in the cerebrum across individuals and is easily detected with conventional brain imaging techniques. Damage to the precentral gyrus, caused by a stroke, for example, can lead to motor deficits on the opposite side of the body. In the cerebellum, foldings are not traditionally referred to as sulci or gyri, but rather as **folia.** A group of folia are further classified into lobes and lobules (lobules are simply smaller lobes). Foldings in the cerebellum of primates, cats, dogs, and rodents make up 10 lobes that have different Latin names describing their shape and location. However, this naming system may be complicated. An easier way to refer to the different lobes has been to use Roman numerals from one to ten (I–X), spanning the cerebellum from front to back (on a rostro-caudal axis) (Figure 3.2). Within each lobe, however, the lobules are not numerically labeled any further.

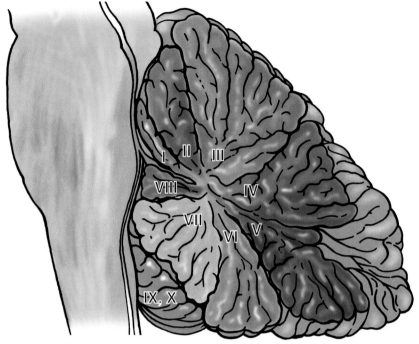

Figure 3.2 The ten lobes of the cerebellum are shown, using the Roman numerals from one to ten (I–X), spanning the cerebellum from front to back (on a rostro-caudal axis).

GENERAL ANATOMY AND PHYSIOLOGY OF THE CEREBELLUM

The cerebellum shares many similarities with the rest of the brain, including the outside general appearance such as folding and divisions (fissures) along the sagittal and horizontal planes. Therefore, the cerebellum is divided into two major hemispheres on either side of the sagittal plane, with a midline structure called the **vermis** (Latin, *worm*) interposed between these two hemispheres. Horizontally, one major fissure separates the anterior from the posterior cerebellum. A sagittal cut through the cerebellum also reveals the typical separation of gray matter from white matter, with the gray matter occupying the most superficial layers of the cerebellum and thus forming the cerebellar cor-

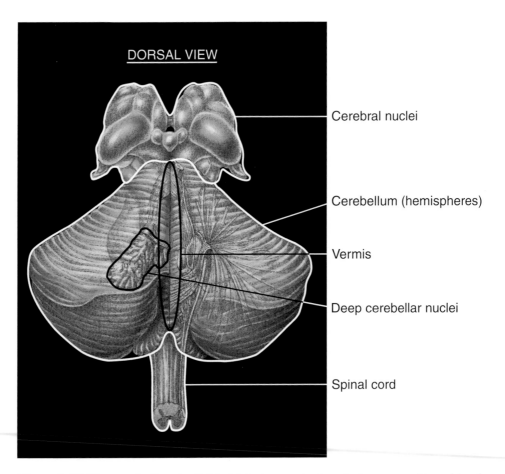

DORSAL VIEW

Cerebral nuclei

Cerebellum (hemispheres)

Vermis

Deep cerebellar nuclei

Spinal cord

Figure 3.3A The cerebellum is divided into two major hemispheres on either side of the midsagittal plane along a structure between the lobes called the vermis. The anterior and posterior aspects of the cerebellum are divided horizontally along a deep fissure (not shown here).

tex (Figure 3.3). Also deep within the cerebellum are three pairs of nuclei called the deep cerebellar nuclei. These nuclei contain cell bodies of neurons (therefore also gray matter) that project their axons out of the cerebellum to other parts of the brain or to the spinal cord.

From the deep cerebellar nuclei, projections descending to the spinal cord and those ascending toward the brain are mostly

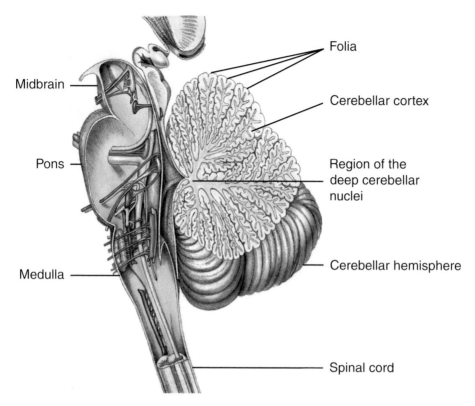

Folia

Cerebellar cortex

Midbrain

Region of the
deep cerebellar
nuclei

Pons

Cerebellar hemisphere

Medulla

Spinal cord

Figure 3.3B This sagittal view of the cerebellum shows the typical separation of gray matter and white matter.

indirectly relayed through different nuclei in the brain stem. As a result, the deep cerebellar nuclei are considered the final output of the cerebellum. Neurons in the deep cerebellar nuclei coordinate motor movements by adjusting descending voluntary motor commands from the brain before they reach the spinal cord or the brain stem for execution. Damage to these nuclei leads to motor deficits such as imbalance and uncoordinated body movement or gaze. The deep cerebellar nuclei are located below the cerebellar cortex, embedded within in a thick layer of myelinated axons (white matter). These axons project to and from the cerebellar cortex. The final output from the deep cerebellar nuclei exits the cerebellum through "bridges" that connect the whole cerebellum with the rest of the nervous system. Recall that the cerebellum occupies the dorso-caudal aspect of the brain and lies just dorsal to the brain stem. In fact, the cerebellum is connected to the rest of the brain only through these three "bridges" on either side of the sagittal plane. These bridges are called the **cerebellar peduncles**. All axons entering or exiting the cerebellum must pass through these cerebellar peduncles. If the peduncles are damaged (e.g., because of a growing tumor in the brain stem close to the peduncles), cerebellar deficits will clearly manifest on the side of the body **ipsilateral** to the damage.

4 The Medulla and Pons

Earlier it was emphasized that there is no clear demarcation line between the central and peripheral nervous systems. In other words, these two systems are not completely separate anatomically. Similarly, within the central nervous system, the brain and the spinal cord are not separated by a clear division, although the medulla is considered the boundary between these two structures. In support of this argument, the medulla shares some features with both the spinal cord and the other brain stem structures. The medulla is a good example to illustrate a recurrent theme—the role that the hindbrain as a whole plays. This role is essentially to mediate between the brain and the spinal cord by receiving ascending input from the latter and processing some of this information before it impacts the brain. The hindbrain also processes, coordinates, and modulates descending brain signals before they are carried out at different spinal cord levels.

GENERAL ANATOMY AND PHYSIOLOGY OF THE MEDULLA

A cross-section through the medulla reveals general anatomical organization comparable to the spinal cord but not exactly alike. For example, the size of the cross-section, specifically through the caudal medulla, is roughly similar to the size

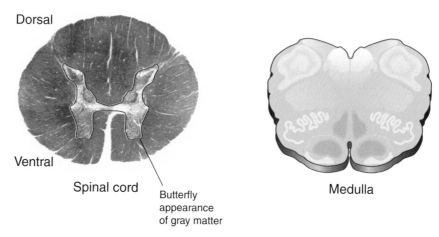

Dorsal

Ventral

Spinal cord

Butterfly
appearance
of gray matter

Medulla

Figure 4.1 A horizontal section of the spinal cord is shown with the "butterfly" appearance of the gray matter compared with a horizontal section of the medulla that shows the more complex appearance of the gray matter there.

of a cross-section through the cervical spinal cord. However, the organization of white versus gray areas differs markedly from that in the spinal cord: The typical butterfly appearance of the spinal cord in cross-section is no longer preserved (Figure 4.1). Instead, a different organization of nuclei in the medulla adds an additional level of complexity in relating anatomy to physiology. Recall that in the spinal cord, the gray matter forms bilateral horns referred to as the **dorsal horns** and the **ventral horns**. The dorsal horns contain cell bodies that mostly receive sensory information related to peripheral events. Accordingly, damage to the dorsal part of the spinal cord may lead to loss of sensation due to the death of dorsal horn neurons. Similarly, selective damage to ventral horn neurons (e.g., due to trauma or neurological disease such as **poliomyelitis**) may lead to the death of motor neurons that send motor commands to peripheral muscles, resulting in **paralysis**. At the level of the medulla, however, such a strict rule of separation between motor and sensory regions is not clearly noticeable; horn structures are not as easily identified as they are in the spinal cord, and nuclei that serve sensory or motor functions can be located either in the dorsal or the ventral aspects of the medulla. In fact, several nuclei can be diffusely located in a complicated network referred to as the **reticular formation**. Neuronal ensembles in the reticular formation coordinate many reflexes generating simple motor patterns of the head and the neck region (e.g., chewing and eye movements), as well as many other patterns that don't require brain intervention, including certain aspects of feeling, as will be discussed below.

■ **Learn more about poliomyelitis** Search the Internet for *polio* or *poliomyelitis.*

NUCLEI AND CRANIAL NERVES OF THE MEDULLA

In addition to being an integral part of the hindbrain, the medulla is considered a part of the brain stem as well. Within the brain stem, many nerves mediate sensory and motor functions for the head and neck region. Because these nerves emerge from the **cranium**, they are referred to as the cranial nerves, and some serve as modified spinal cord nerves. In the spinal cord, peripheral nerves emerge from, or enter, the spinal cord through dorsal or ventral roots to mediate either sensory or motor functions, respectively. As a result, selective damage to these peripheral roots making up the peripheral nerves may result in either sensory or motor deficits. On the other hand, cranial nerves do not typically share this clear sensory versus motor attribute, if compared only based on their dorsal versus ventral location in the brain stem. In addition, some cranial nerves may have sensory or motor components, while other individual nerves may be both sensory and motor, and may also include control of **autonomic** functions. Twelve cranial nerves have been identified and numbered using

Roman numerals I–XII and the initials CN; for example, the seventh cranial nerve (the facial nerve) is designated as CN VII. Three of these nerves (IX, X, and XII) and their nuclei are located at the medullary level and control mainly motor functions related to mastication, swallowing, lip and tongue movements, and respiratory reflexes. In addition, these nerves transmit sensations from the upper airway, as well as taste from a portion of the tongue and oral cavity.

THE MEDULLA IS PART OF BOTH THE HINDBRAIN AND THE BRAIN STEM

Along with the intricacy and diversity in the functions of the medulla, there is an additional level of complexity in the nomenclature associated with this structure. The brain stem includes the medulla, the pons, and other structures, whereas the hindbrain, which we're mostly concerned with here, also includes the medulla and the pons, as well as the cerebellum. Therefore, depending on the nomenclature used, one has to be careful in ascribing functions to these distinct brain regions. For example, if one assumes that the medulla mediates movement patterns during chewing, this does not necessarily mean that the entire brain stem or the entire hindbrain plays a role in mastication. Accordingly, although the hindbrain and the brain stem share certain anatomical structures, it would be best to discuss each part within these structures separately to avoid confusion.

THE PONS

The pons (*bridge*) is located rostral to the medulla. Neurons in **pontine nuclei** relay (connect) between neurons in brain stem nuclei and those in the cerebellum. The pontine nuclei receive neuronal signals descending from other brain areas, mainly the cerebral cortex, and relay this information, which is important for the control of voluntary eye movements and the coordination of eye and body movements, to the cerebellum.

GENERAL ARCHITECTURE OF THE PONS

Neurons in the pons aggregate in nuclei (which is why they are referred to as pontine nuclei) to form a dense and relatively large zone of gray matter on either side of the midline. Recall that nuclei are formed by cell bodies of neurons that have comparable functions or process similar input grouped together. These pontine nuclei are interrupted by passing myelinated axons traveling in the rostro-caudal direction. As a result, the pons exhibits a particular histological landmark of dense gray matter that looks like a "grid" interrupted by white matter.

Because the cerebellum coordinates movement commands initiated in the motor cortex, it should not come as a surprise that axons in the pons carry information from the cerebral motor cortex (actually from the contralateral cerebral hemispheres). These neurons synapse on other neurons in the pontine nuclei that then send their axons across the midline to the contralateral side of the cerebellum. In the cerebellum, these axonal endings may synapse either on neurons in the cerebellar nuclei or on other types of cerebellar neurons (see Chapters 5 and 6). Within the cerebellum, processing of input related to voluntary movement involves many intricate internal pathways, until eventually this input reaches the main decoder of information in the cerebellum: **Purkinje cells**. Purkinje cell networks (input from climbing and parallel fibers and projections mainly back to neurons in the deep cerebellar nuclei) will be reviewed again later. Through the cerebellar nuclei, Purkinje cell output feeds back to neurons of the contralateral cerebral motor cortex via another relay in the thalamus. Thus, a motor scheme may be planned in the motor cortex, but is rarely executed before it is first coordinated by the cerebellum, maestro of the orchestra.

Furthermore, following the paths of connections between the motor cortex and the cerebellum, one side of the cerebellar hemisphere controls the ipsilateral side of the body's fine movements, in contrast to the cerebral motor cortex, whose hemispheres control the contralateral sides. As a consequence, one should expect that severe damage to one side of the cerebellum might result in loss of coordination (**ataxia**) on that same side of the body.

5 | Purkinje Cells

The cerebellum is home to several different types of neurons. Purkinje cells stand out from other neurons in the cerebellum and occupy center stage in cerebellar physiology in the way they look and how they "sound" when their electrical activity is recorded. First identified by Czech physiologist Jan Evangelista Purkinje (1787–1869), these neurons were also anatomically described later by Spanish scientist Santiago Ramón y Cajal in many of his meticulous illustrations that, even today, remain accurate and a great testimony to the elegance of neuroscience histology (see "Santiago Ramón y Cajal" box). Purkinje was born in Bohemia (part of what is now the Czech Republic) in 1787. He graduated from the University of Prague in 1819 with a degree in medicine. After publishing his doctoral dissertation on vision, he was appointed professor of physiology at the University of Prague. Purkinje created the world's first department of physiology at the University of Breslau, Prussia, in 1839 and the first official physiological laboratory, known as the Physiological Institute, in 1842. He is best known for his discovery, in 1837, of Purkinje cells—large nerve cells with many branching extensions found in the cortex of the cerebellum.

■ **Learn more about Purkinje cells** Search the Internet for *Jan Evangelista Purkinje* or *Purkinje cell.*

Purkinje cells underlie the main function of the cerebellum—that is, "orchestrating" brain functions. This chapter discusses the basic functions of Purkinje cells. Chapter 6 discusses neurological deficits associated with the cerebellum, and Chapter 7 provides a more detailed discussion of cerebellar circuitry and function.

PURKINJE CELL LAYER

Along any sagittal cut through the cerebellum, a peculiar line of neuronal cell bodies appears along the superficial edges of the cerebellar lobes. These cell bodies belong to Purkinje cells and they're easy to distinguish (Figure 5.1). They are relatively larger than other neurons, and all are aligned with their dendrites facing the surface while their axons project inward, deep into the cerebellar white matter. These axons eventually synapse on dendrites of neurons located in the deep cerebellar nuclei. In fact, the Purkinje cell layer is so prominent that it is considered a landmark of the cerebellar cortex.

Santiago Ramón y Cajal (1852–1934)

Spanish scientist Santiago Ramón y Cajal suspected that the nervous system was made up of many individual cells that were separate from each other. To test his hypothesis, he used a special dye that was designed by Italian scientist Camillo Golgi to stain various cells of the brain. He then examined the brain cells with a microscope, and sketched what he saw in a series of elaborate pictures. Using this approach, Cajal demonstrated in 1887 that the brain is indeed made up of individual brain cells, which are called neurons. Ramón y Cajal and Golgi shared the Nobel Prize in Physiology or Medicine in 1906 for their important work on the structures of the nervous system.

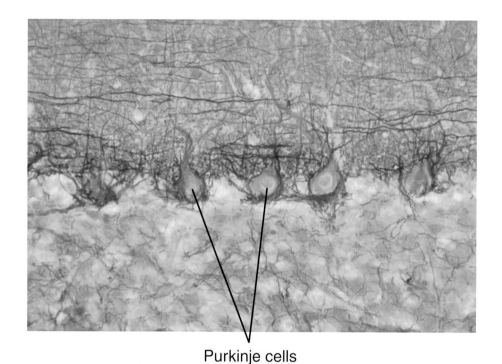

Purkinje cells

Figure 5.1 This light micrograph shows a row of Purkinje nerve cells (note large cell bodies) magnified 100 times.

PURKINJE CELL INPUT

Each Purkinje cell receives two main types of input on its dendrites. The axons that synapse on Purkinje cells originate mainly from climbing fibers or parallel fibers. Each Purkinje cell receives projections from only one climbing fiber, but may receive synapses from as many as 25,000 parallel fibers. To better understand cerebellar circuitry, it helps to understand the nomenclature of cerebellar histology.

For example, a **climbing fiber** refers to a neuron whose cell body is in a nucleus in the brain stem called the inferior olive. Axons from these fibers "climb" toward a Purkinje cell to form intricate and extensive synapses with its dendritic column. On the other hand, **parallel fibers** form parallel lines that run horizontally across the cerebellum, contacting many Purkinje cells on their way. When an **action potential** travels along a climbing fiber, it always results in an action potential in the Purkinje cell on which it synapses. Accordingly, the synapse between a climbing fiber and a Purkinje cell is the most "secure" and powerful of the entire nervous system. In fact, electrophysiological recordings close to these synapses reveal action potentials of the largest amplitude in the whole nervous system. The action potential is also remarkable because it has one big spike followed by smaller ones called secondary spikes. In other words, every time a climbing fiber is activated, a powerful **complex spike** is generated in the Purkinje cell. If the recording setup is connected to an audio amplifier, the sound elicited by the complex spike can be recognized easily even without looking at the oscilloscope. In contrast, activation of a parallel fiber rarely elicits an action potential in a Purkinje cell, unless enough parallel fibers synapse on a given Purkinje cell at a particular moment. An action potential generated in this manner results in a less powerful action potential called a **simple spike**. Recording near a Purkinje cell **dendritic tree** shows that Purkinje cells are always spontaneously active, firing on average one complex spike every second and approximately 20–60 simple spikes per second. Recordings also sound as though Purkinje cells go quiet immediately after a complex spike, only to resume rapid simple spike activity afterward. This brief but variable period of quiet (e.g., ranging from a few hundred milliseconds to seconds) is thought to underlie the phenomenon of cerebellar learning. The idea of comparing brains to computers is based on this observation, as discussed in Chapters 7 and 8.

PURKINJE CELL OUTPUT

Axons of Purkinje cells project away from the surface and deep into nuclei inside the white matter of the cerebellum. These

nuclei span the sagittal plane of the cerebellum, with three nuclei on each side. Typically, Purkinje cells to the left of the midline project to the left group of nuclei and those on the right
project to the same-side nuclei. An exception to this scenario is
the direct projection of Purkinje cell axons beyond the deep
cerebellar nuclei and even outside of the cerebellum toward the
brain stem, reaching a group of neurons called the vestibular
nuclei. All Purkinje cells at their terminal endings release a
neurotransmitter that causes inhibition in the postsynaptic
neuron. This neurotransmitter, called GABA (gamma amino-
butyric acid), and its receptors are widely distributed in other
parts of the brain and the spinal cord. As a result, every time a
Purkinje cell is activated, a neuron lying in the deep cerebellar
nuclei (on the same side and often along the same sagittal
plane) is inhibited.

PURKINJE CELLS REMEMBER AND LEARN

The ability to learn is a distinctive feature of the Purkinje cell
input-output network and the different patterns of activity generated in its cell body and axon, and communicated to target
neurons in the cerebellar nuclei. Defined broadly, learning requires attention, storage, and retrieval. A relevant example
would be a reader learning new information by reading a book.
In order to learn, one must first pay attention. Distractions
caused by loud sounds, people talking, or music, or even motor
distractions such as eating or walking while reading, all reduce
attention, and thus reduce learning ability (this is probably why
sitting in a quiet library is conducive to better learning). Reading attentively is necessary for learning, but it is not enough.

Understanding is also equally important (perhaps this is why
one often needs to read the same paragraph again and again to
understand and filter the material instead of simply memoriz-
ing it). And finally, retrieval of stored information is necessary

for learning. Learning can be measured with several types of tests. Tests may come in academic forms (written or verbal), or in social "life events" forms (testing one's ability to solve daily problems or tests of motor behavior, such as during a gymnastics competition).

Adaptation underlies the survival of all beings, and adaptation includes learning. Purkinje cells can learn, too: They can store information of past important events, mainly motor events, and thus help us adapt. For example, tripping over a crack while walking is a signal of error in motor performance that needs to be corrected immediately to protect us from falling and being injured. Purkinje cells seem to receive information related to error messages, presumably to condition the motor system to perform better under similar conditions in the future (avoid tripping by better responding to sudden obstructions). As the Purkinje cell is regularly firing simple spikes at a high **frequency** (i.e., events per second measured in Hertz, or Hz, units; e.g., 2 events per second = 2 Hz) with an intermittent but somewhat predictable complex spike pattern (1–2 Hz), an error in motor performance (tripping or losing balance) will activate neurons in the inferior olive, which will then activate its axons along climbing fibers synapsing on Purkinje cells, resulting in an increase in complex spike firing. A change in the pattern of complex spike firing acts as a "teacher" to simple spikes, causing a reduction in their rate of firing. This will be explained in more detail within the general context of cerebellar circuitry in Chapters 6 and 7. However, a good example is to imagine a classroom with 100 students (parallel fibers) being loud and a teacher (one climbing fiber) then being annoyed and suddenly asking the students to quiet down. This is like a transient learning episode for students, after which they may resume their loud chatter, unless the teacher steps in again and again, until the rumbling stops and the "error" is corrected.

This truly marvelous property of the Purkinje cell has inspired many computer engineers, mathematicians, and artificial intelligence experts to create models for machines that could one day learn efficiently based on these seemingly strictly biological properties. The bases for these properties and ideas will be discussed in the next two chapters.

6 Functions and Malfunctions of the Hindbrain

How is the function of an area in the brain determined? Traditionally, it is the absence of a function, or, in other words, a neurological deficit that directly links a damaged brain area to its function (which can often be verified post-mortem by autopsy). For example, injury to the front of the head that damages the frontal lobe of the brain is linked to intellectual deficits. Similarly, a stroke in the cerebral cortex may cause either paralysis or sensory loss, depending on the stroke's location within a few millimeters of the **central sulcus**. Thus, in humans, the frontal lobe has been assigned a role in cognitive ability. Similarly, the roles of the cerebellum and other hindbrain structures were initially established. Some of the functions that have been attributed to the cerebellum and the medulla are explained here.

NEUROLOGIST SIR GORDON HOLMES

During World War I (1914–1918), neurologist Gordon Morgan Holmes treated many soldiers with head wounds and brain injuries and kept accurate records of the cases he encountered. Of particular interest to Holmes were injuries to the lower back of the brain (including the cerebellum) that caused patients to experience severe motor deficits (but not paralysis). Holmes was born in Ireland. He was a shy, solitary child who qualified at Trinity College Dublin in 1899. In Dublin, he was awarded the Stewart Scholarship, which financed a visit to Frankfurt, where he studied comparative anatomy with Dr. Carl Weigert (1845–1904, famous mostly for his method of staining myelinated axons) and Ludwig Edinger (1855–1918, considered the founder of modern neuroanatomy). When he returned from Germany in 1902, Holmes quickly mastered clinical work and brought it to a state of nearly scientific perfection. His exhaustive yet rapid examination routine was unprecedented. He had just joined the staff of a Red Cross hospital behind the frontline in France, when myopia (nearsightedness) hampered his vision and his ability to serve in the war. Because of his impressive work, the War Office was persuaded to keep him on. Holmes studied specifically the cerebellum and the **visual cortex**, which were exposed in soldiers whose helmets did not cover the backs of their heads. He treated up to 300 wounded men daily, and later published 18 wartime papers, many written under arduous conditions.

German neurologist Hermann Munk described Holmes:

> With his tall, powerful frame and his hawk-like eyes under beetling brows and spectacles, he intimidated candidates for the College Membership [examination] until they found that direct answers to direct questions brought out the kindliness for which he was known to his intimate friends.

British neurologist Macdonald Critchley said:

> The story goes that whenever Holmes and [Kinnier] Wilson made their respective rounds in Queen Square, each with his own retinue of doctors of all ranks, and they met in the passageways, neither of them would budge to make way for the other party.

Lengthy blockages ensued. . . . Many neurologists treasured the memories of their apprenticeship to one of the giants of neurology, and to a staunch, fundamentally warm-hearted counsellor and guide. In his profession, as in his garden, Holmes planted seeds for the profit and wonderment of generations to come.

Holmes developed the foundations of modern neurological examination, and his writings were comprehensive and regarded as authoritative in the field of neurology, specifically those related to cerebellar motor deficits such as ataxia.

■ **Learn more about Gordon Morgan Holmes** Search the Internet for *Gordon Morgan Holmes.*

CEREBELLAR ATAXIA

In clinical terms, a *deficit* refers to loss of function or poor performance. The motor deficit classically related to cerebellar damage is ataxia. *Ataxia* is actually a compound word of the letter "a" and the word *taxia.* The prefix "a" denotes inability or "without," and *taxia* refers to order. Therefore, *ataxia* is defined as the inability to perform accurate (orderly) voluntary movements. Ataxia is a motor deficit that is not life-threatening but that affects the execution of fine movements. Therefore, many social aspects of life, including walking gracefully or performing rapid alternating movements accurately (e.g., playing piano or doing gymnastics), are affected in people with ataxia. Ataxia may result from damage to other parts of the brain as well, such as the basal ganglia, as is the case in Parkinson's disease, which is different from cerebellar ataxia. Cerebellar ataxia does not result in the inability to move, or paralysis; rather, it causes poor movement: Patients will eventually get from one place to another, but they tend to have a wobbly gait. A common test for cerebellar ataxia is the "finger-to-nose" test. In this test, the patient is asked to stretch out his or her arm and then move the index finger to touch the tip of the nose. A cerebellar deficit will manifest itself as a zigzag path traced by the index finger attempting to reach the nose rather than a straight line back and forth. In this same test, a patient with a cerebellar deficit will be slow to initiate the movement. One explanation for these deficits is the inability of the cerebellum to correct for errors in movement, occurring as a result of the cognitive will or the desire to accomplish these motor tasks. Laboratory experiments support the hypothesis that the cerebellum is fundamental for storing and retrieving information about motor events and thus motor learning.

The cerebellum may have other functions in addition to motor learning. Recent, more thorough, examination of patients with cerebellar ataxia has revealed that some of these patients might also suffer from problems in speech and cognitive problems such as schizophrenia. These observations have yet to be corroborated with more solid experimental evidence.

■ **Learn more about Parkinson's disease** Search the Internet for *Parkinson's disease* or *ataxia*.

VISION AND EYE AND HEAD MOVEMENTS

What most of us take for granted, our sight, is really a series of mechanisms with overwhelming complexities. First, there is the sensation of visual information. Vision starts from the moment the light reflecting off an object in our visual field impacts on the **retina** of the eye, transforming light energy into electrical activity in neurons. This neuronal activity subsequently travels in discrete neuronal pathways to specialized brain regions where visual perception of the stimulus is thought to finally emerge. Consider the following example related to me by my friend Tracey:

I drove past the house I grew up in and stopped to look at the large oak tree that dominated the front yard. It's still as big as I remember, with those massive limbs stretching out over part of the street. We used to climb up and sit on them, watching as cars passed underneath. Its branches near the base were unusually large, twisted, and stretched up high, providing ample shade and a multitude of leaves that had to be raked up every fall. It gave me comfort to know that the old solid tree was still there.

Tracey followed those leaves with her eyes wide open, without moving her head, as the tree was far enough from where she was standing. Tracey is merely contemplating a visual cue. Yet even this simple exercise triggered a powerful emotion in her, and that cannot be regarded as a simple act of transformation of light into a "photograph" in her mind with no meaning. Eye movements are only one aspect of vision, which is a complex phenomenon that is well connected to brain structures involved in cognition and emotion, a window to our perception.

Now let's look at how a moving target triggers neuronal activity. Never mind a tranquil scene such as the oak tree in the distance. . . . Imagine now a leopard hunting a gazelle. The leopard will typically hide behind some vegetation, preferably camouflaged, and track the gazelle constantly with its wary eyes. As a moving target, the gazelle poses a challenge to the leopard. Constantly moving around across the field, foraging for food, the gazelle reaches the "corner" of the leopard's eye. To keep its prey well within its visual field when eye movements alone are not enough, the leopard moves its head to accommodate its visual field and keep the gazelle in sight. This is one example when the target is in motion but the body is stationary.

A third example involves a stationary visual cue seen while the body or the head itself is in motion. Focus on one object in front of you, located at any distance, and rotate your head slowly and smoothly to the left while keeping an eye on the visual target (keep it in focus). It may be hard to realize that, in doing so, your eyes have simultaneously moved with your head rotation and with the same velocity, although in the opposite direction (to the right). To better illustrate this point, close your eyes and rotate your head to the left while trying not to move your eyes, then open your eyes and you'll definitely realize that your target is no longer within focus. As a rule, stationary objects stay in focus while the head is moving only if the eyes compensate by moving with the exact opposite velocity and in the opposite direction relative to head movements. Clearly, then, keeping visual targets in focus requires fine motor control and intricate coordination of the sensory visual cues (visual sensory input) with the muscles controlling eye movements (visual motor output). These pathways can be traced along the path of input from the retina of the eye, to many brain structures and brain stem nuclei (including pontine nuclei), some ultimately reaching the cerebral sensory visual cortex for visual perception, and project back to eye muscles as motor output from the cerebral motor cortex. Superimposed upon this schematic system (sensory input and motor output) is the role that the cerebellum plays in the coordination of eye movements with head and body rotations. Specific areas in the cerebellum are specialized for the coordination of this specific motor behavior before the final output reaches the final effector organs (eye muscles). Therefore, damage to specific brain stem nuclei or cerebellar structures can lead to impairment in eye movements, but not necessarily visual impairment. In other words, during a routine neurological exam patients might not show signs of decreased visual acuity, but may instead fail to follow the examiner's finger as it slowly travels across their visual field (if head movements are not allowed). In addition to these details, in most of the eye movement examples described above, it is estimated that both eyes move synchronously (in the same direction and with the same velocity), and that, therefore, an intricate neuronal network must somehow connect these muscle groups in both eyes. In fact, such a network is located within the brain stem, and controls **antagonistic muscle** groups— that is, if both eyes rotate to the right, the right eye is pulled by its right muscle (lateral rectus) and the left eye is pulled by its right muscle as well (medial rectus), simultaneously.

PHYSIOLOGICAL BASIS FOR CEREBELLAR FUNCTION

It was already stated that Purkinje cells occupy center stage in the function of the cerebellum. Briefly stated, Purkinje cells inhibit downstream neurons in the deep cerebellar nuclei by sending inhibitory messages on a regular basis (1–2 Hz), which accelerates after an error in motor performance or an abrupt need to correct for an action that will happen (or one that is already taking place). Neurons in the deep cerebellar nuclei then send their axons to many areas in the brain stem to control (adjust) movement. Without this second-to-second (perhaps even faster) monitoring and adjustment of motor performance by "keeping an eye" on brain stem neurons that mediate movements, cerebellar deficits emerge (ataxia, imbalance, incoordination). The concept of correcting for errors in execution can also explain speech problems associated with cerebellar damage. Such problems can be explained by the inability to correct for errors in articulation of speech, or the inability to correctly "put a thought into words." Within this context, it becomes easy to imagine the cerebellum as a maestro conducting a symphony; surely, the band can still perform without the conductor (or to use clinical analogy, the muscles can still move), but the music will be out of tune (ataxic). This explains why many of the subtle cerebellar deficits have remained undetected until careful clinical examination later revealed these cognitive deficits, especially with regard to schizophrenia. Future studies may shed more light on other functions of the cerebellum, such as the coordination of emotions and perhaps even the body's response to stress.

LATERAL MEDULLARY SYNDROME

Based on the information given above regarding the anatomy and physiology of the medulla, certain deficits associated with medullary damage can be accurately predicted. One such group of symptoms is referred to as the lateral medullary syndrome, also called **Wallenberg syndrome**. It was first described in 1808 by Gaspard Viesseux (1746–1814), who rendered an exact description of his own disease at a meeting in Geneva, Switzerland. The case was published in 1811 after Viesseux had visited London and reported his disease there. It was Adolf Wallenberg who first described the condition clinically in 1895 and later verified it by autopsy in 1901. The disorder is generally caused by a blockage in an **artery** (either the vertebral artery or a branch of the posterior inferior cerebellar artery) that supplies part of the medulla and the inferior aspect of the cerebellum, thus causing irreversible neuronal degeneration of cell bodies or nerves in that region. When patients suffer from this neurological disorder, they typically have swallowing difficulties and hoarseness, resulting from paralysis of a portion of the vocal cord; loss of taste; and some paralysis of the facial muscles. Most of these signs can be explained based on an understanding of cranial nerve anatomy and the nuclei of cranial nerves within the medulla. However, symptoms may also include dizziness and loss of pain or temperature sensitivity. Therefore, this condition can be compared to the deficits associated with spinal cord lesions. For example, following injury to the dorsal aspect of the spinal cord, loss of fine touch is often reported. This could be due to interruption of neuronal signals traveling in white matter axons forming **tracts** ascending in the dorsal spinal cord. In the lateral medullary syndrome, loss of pain could be caused by interruption of the equivalent tracts for pain ascending through the medulla. In fact, this peculiar neurological condition is usually characterized by loss of pain and temperature sensation on one side of the body above the neck, and similar deficits below the neck on the contralateral side. This point could well be explained on the basis of anatomical differences in the paths taken by the different tracts mediating touch and pain, a point that is beyond the scope of our discussion in appreciating the basic functions of the medulla. Note, furthermore, that individuals with Wallenberg syndrome frequently report an unsettling tilt of their environment, which affects their balance, presumably due to damage to the eighth cranial nerve (CN VIII). Treatment for Wallenberg syndrome is symptomatic—that is, individual deficits are treated separately. For example, if swallowing is greatly impaired, a feeding tube may be used. Physical therapy and intraoral appliances may aid speech and swallowing, and in some cases, medication can reduce or eliminate pain. In fact, for some patients, symptoms go away after a few weeks of treatment, not because the disease has been cured, but because the central nervous system has an impressive ability to compensate sometimes for deficits.

■ **Learn more about Wallenberg syndrome** Search the Internet for *Wallenberg syndrome.*

7 | Of Brains and Machines

It has been stated before that a section of the cerebellum of a fish viewed under a microscope is almost indistinguishable from that of a rat or a man. In fact, the cellular organization of the cerebellum is so spectacularly regular that it has been tempting for many biologists and mathematicians to compare it to a machine. This rather novel approach of looking at a living tissue from a mechanistic point of view was controversial at first, but soon gathered significant support and ultimately credibility from the scientific community. What initially fueled the comparison was a basic physiological principle underlying the function of the nervous system, which will be discussed below. Although the basic physiology of neuronal cells has been discussed already, it is worth reviewing some basic neuroscience facts in comparison to computers to further highlight this description.

THE BINARY CODE IN COMPUTERS

The 1940s saw the birth of computers. Of course, these computers were primitive compared to their modern-day counterparts. Although their storage and computing capacities were far inferior to what we know today, they were nevertheless a revolution in the way people viewed machines. Suddenly, daring and novel questions such as "Can a machine really think?" were being seriously considered from philosophical, biological, engineering, and mathematical points of view. From a historical perspective, it is worthwhile to illustrate here the controversy over the new machines, which was lucidly reflected by Edmund Callis Berkeley in 1949 in his book *Giant Brains, or Machines that Think*:

> Recently, there has been a good deal of news about strange giant machines that can handle information with fast speed and skill. They calculate and they reason. Some of them are cleverer than others—able to do more kinds of problems. Some are extremely fast: one of them does 5,000 additions a second for hours or days . . . these machines are similar to what a brain would be if it were made of hardware and wire instead of flesh and nerves. It is therefore natural to call these machines mechanical brains. Also, since their powers are like those of a giant, we may call them giant brains.
>
> Several giant mechanical brains are now at work finding out answers never before known. Two are in Cambridge, Massachusetts; one is at Massachusetts Institute of Technology, and one at Harvard University. Two are in Aberdeen, Maryland, at the Army's Ballistic Research Laboratories. These four machines were finished in the period 1942 to 1946. More giant brains are being constructed. Can we say that these machines really think? What do we mean by thinking, and how does the human brain think?

■ **Learn more about Edmund Callis Berkeley** Search the Internet for *Edmund Callis Berkeley*.

Evidently, this debate sparked not only discussions about the possible intellectual abilities of machines, but also far-reaching discussions concerning the whole process of human intelligence. Although considerable progress has been made in our understanding of human mental processes such as learning and memory, especially during the latter part of the 20th century, a fundamental question such as "How does the human brain think?" remains largely unanswered. What Edmund Berkeley judicially pointed out in 1949 still holds true today, "If you ask a scientist how flesh and blood in a human brain can think, he will talk to you a little about nerves and about electrical and chemical changes, but he will not be able to tell you very much about how we add 2 and 3 and make 5." At any rate, what concerns us more here are questions relevant to brains rather than machines in general. Therefore, suffice it to say that computers operate on a basic electronic principle. All the internal electrical cables and hardware that make up a computer are built in a way either to allow a current to flow in a certain circuitry or not. An electric cable within a computer, at any given instant, is said to be either "active" or "not active," i.e., either "on" or "off." When it's "on," it allows only one type and one amount of current. As a result, if we were to monitor the electrical activity within a given electric cable in an operating computer, assigning the symbol 1 for "on" and 0 for "off," the report we generate will logically only contain 1 and/or 0 symbols. For example, a record of electrical activity in a given cable could then look like the following: 1, 1, 1, 0, 1, 0, 0, 1, 0 . . . or 0, 1, 0, 1, 0, 0 . . . , etc. These symbols **encode** a message communicated by electrical transmission through the electrical cable. Because only two symbols are used, it is referred to as **binary code** (*bi* means "two"). Without going into great mathematical detail, we would assume here that, even using only two symbols, the binary code is indeed enough to encode an infinite number of messages. But how can this fundamental, yet very simple, principle that underlies the magnificent capabilities of these giant brains called computers be related to our biological living brains?

THE BINARY CODE IN THE LIVING BRAIN

Neurons are the basic cellular units of the nervous system. Neurons are also supported by glial cells that are actually more abundant than neurons. But the main physiological functions of the nervous system are thought to be mediated essentially by neurons. With its extensions (dendrites and axon), a neuron is capable of receiving (through its dendrites) and sending (through its axon) information to other neurons or **effector organs** (e.g., a muscle or a gland). Contacts between neurons are also called **synaptic zones**, and communication between neurons is relayed through the synapse. Generally, in a synapse a chemical (in fact, a neurotransmitter) is released from one axonal terminal to signal a message to another neuron. Usually, the neurotransmitter released from one axonal terminal would act on specific **receptors** expressed on the dendritic extensions of the other neuron (but other modes of synapse are also possible, such as an axo-axonic synapse, or a synapse between one axon and another). Exceptions to this general rule include signaling of a neuron to an effector organ rather than to another neuron; or two neurons may have their cellular membranes in direct contact and may form tight membrane junctions, whereby electric synapses flow directly from one neuron to the other without the need for a chemical release to relay the signal.

In any case, if the message from the first neuron is strong enough, the second downstream neuron will then initiate another message. In this case, the signal is said to have crossed the threshold of excitation of the second neuron. It is important to note here that the incoming message from the first neuron is not always excitatory; it could also be inhibitory, depending on the neurotransmitter released from the axonal terminal and the type of receptors expressed on the other side of the synapse (postsynaptic). Therefore, the second neuron downstream from the first one (postsynaptic) will first compute the sum of all incoming messages from different sources before judging the final outcome. However, once the sum of the incoming message(s) reaches or exceeds the set threshold, the signal initiated in the second neuron obeys the law of **all-or-none response**. That is, the signal traveling along an axon has invariable characteristics that are unique to that particular neuron at any given time and regardless of the message intensity. For example, if a neuron receives a moderate message that crossed its threshold of activation, that same neuron may send one or two signals down its own axon. But if the message is more intense, or acts for a longer period at the synapse, it will then trigger that same neuron to fire a larger number of signals, perhaps also within a shorter period of time (at a high frequency). But in both cases, the signals that travel down the axon are unique to each neuron. This is called a unitary response, and the unit in question is called an action potential. Action potentials propagate at a given speed depending on several neuronal properties (including axonal diameter and degree of myelination). The velocity at which action potentials travel is normally always the same for a given neuron, unless the axonal properties change due to a disease. For example, **multiple sclerosis** is a disease that damages myelin around axons in the central nervous system, and therefore slows down conduction in these axons. If we compare a neuron to a gun, the threshold for activation to the safety catch, and the action potentials to the bullets, a neuron can be said to "fire" each time a trigger is "pulled" strongly enough to send an action potential (the bullet) shooting down its axon. According to this scenario, a neuron can either fire or not fire, and when it does, it fires the same-sized bullet at the same speed along a fixed path. It becomes clear then that the nervous system also uses a binary code, and that electric cables indeed resemble neuronal axons in many ways, and that computers arguably do mimic the living brain, at least from a functional point of view.

THE CEREBELLUM AS A COMPUTING MACHINE

First and foremost, the cerebellum is a part of the nervous system, and therefore obeys the fundamental rule of binary coding in its intrinsic neuronal circuitry. However, an additional level of similarity exists between the cerebellum and computing machines. A typical computer consists of hardware (built-in electrical wiring underlying the internal circuitry) and a set of rules or commands that makes use of the existing network in a specific manner (software). For example, it is possible to change the operations of a computer by changing the software. A change in operations can also be achieved by changing the hardware instead (i.e., by changing its internal network organization), but that is more difficult to do. This ability to change (by changing the software) is comparable to rapid adaptation to environmental changes in living organisms (except that computers cannot automatically and efficiently adapt and learn, at least not yet). This versatility in function, exemplified in the ability to change the software instead of the hardware, adds tremendous versatility to the functions that a computer can accomplish.

Within the cerebellum, the cellular organization could be compared to hardware, whereby the anatomy is repetitive throughout and very well defined. Basically, two sources of input synapse on Purkinje cells, which then project out of the cerebellar cortex to inhibit neurons in the deep cerebellar nuclei and in other brain stem nuclei as well. This basic hardware, however, is capable of mediating a variety of functions, including the control of motor behavior, in addition to influencing cognitive, sensory, and linguistic functions. With this diversity of functions, the cerebellum could well be relying on different "software" to handle different demands imposed on it by the cerebral cortex, while using only a simple "hardware" circuitry. In fact, in support of this argument, the output from the cerebellum is segregated into different types of bundles depending on the target and the function intended by that specific output. For example, cerebellar output neurons projecting on a particular group of neurons within a specific region or nucleus in the brain or the brain stem may group together, forming a bundle of fibers. This grouping of cerebellar output into bundles of fibers that serve similar functions or project to common brain areas allows the cerebellum to communicate complex and powerful neural symbols, while putting into good use two basic principles in neuroscience: binary code and specialization of function.

To better illustrate the advantages of communicating via bundles of specialized networks, consider a bundle made up of only five neurons. Next, consider hypothetically that these fibers need to communicate a "word" to the target nucleus or downstream neurons. This word is made up of letters of the alphabet, and a letter corresponds to each pattern of firing of these fibers within the bundle. For example, if fiber number one is firing and the other four are silent, this encodes the letter N, whereas if fiber number two is firing while the other four are silent, this is a code for the letter O, etc. (In fact, the number of possible combinations based on this design far exceeds the numbers of the letters of the alphabet.) Therefore, this group of fibers is capable of spelling out the word *no* to the downstream neuron. Therefore, a bundle of only five neurons could theoretically communicate an exceedingly large number of words to the downstream target, based only on binary coding. Experimental research on cerebellar physiology has indeed confirmed the specialization of certain cerebellar bundles in communicating indirectly with regions in the cerebral cortex. Presumably, this not only allows for more powerful and more efficient means of communication, but also apparently permits the cerebellum to communicate with broader and more diffuse areas of the brain simultaneously. It is worth recalling here that the cerebellum is a very dense neuronal mass that actually contains more neurons than found in the entire rest of the brain, and that these neurons help in the coordination of multiple functions; hence, efficiency in physiology (grouping and specialization) and in anatomy (high degree of folding) are features more prominent in the cerebellum than anywhere else in the central nervous system.

FEEDBACK AND THE CONTROL OF ERROR MESSAGES

Let us consider a hypothetical clinical case in which Mr. Clark is a patient who comes into a clinic. Sitting in a chair waiting to be examined, Mr. Clark looks perfectly fine, and nothing abnormal is noticeable. However, when you present Mr. Clark with a pen, he will extend his arm beyond it as he reaches for it, and then he will attempt to correct for this error by swinging his arm in the opposite direction, only to fail to take the pen again. He may continue to try to reach out for the pen until his swing becomes more intense and transforms eventually into violent shaking the closer his hand gets to the pen. Although Mr. Clark's vision is perfectly fine, he obviously suffers from a neurological motor disorder. This motor deficit is referred to as ataxia, or an inability to accomplish an intended motor performance due to inaccurate execution. The underlying causes of ataxia are diverse, but in this case, a clinical examination of Mr. Clark reveals that his ataxia is associated with a tremor toward the end of the intended voluntary motor performance. If Mr. Clark is seated quietly or not attempting to move, no obvious abnormal sign is observed. This specific form of ataxia is referred to as "intention tremor" or "motion tremor." In contrast, another form of ataxia is observed in patients who suffer from Parkinson's disease, whereby the tremor is usually present at rest, and actually diminishes when they make purposeful movements. Parkinsonian tremor is therefore referred to as a "tremor at rest." The underlying causes of these ataxias are different. Although these forms of motor deficits manifest following a neurological disorder that targets the brain, one is specifically linked to the hindbrain in the cerebellum (intention tremor), whereas the other is linked to degeneration of neurons in a different brain region that also controls motor behavior (the basal ganglia).

The functions mediated by different brain regions have traditionally been deduced from clinical signs associated with damage to a particular region. For example, evidence that a specific area in the cerebral cortex controls voluntary movements came following the link established between brain strokes in the motor cortex and motor deficits on the contralateral side of the body. Similarly, links have been established with cerebellar abnormalities in the case of intention tremor (in addition to other motor deficits of cerebellar origin, such as loss of motor coordination and unstable gait). As a result, the coordination of motor movements has been traditionally attributed to the cerebellum. More specifically, intention tremor reflects a failure in correcting for an error in a voluntary movement.

To better illustrate this idea, consider the following example: Suppose you were asked to build a machine to control the heating of an apartment. This machine needs to be automatic, meaning that it needs to operate independently, to control temperature without surveillance. Unlike a gas stove used for cooking that must be operated by an outside agent (the cook), you are not allowed to stay close to the heater to turn it off when it becomes too hot, or to turn it on again when it gets too cold. In fact, you are required to maintain the apartment at a certain temperature, or at least within a certain narrow temperature range, let's say at 70°F ± 2°. On its own, the hypothetical heater will not be able to "sense" the temperature to decide when to turn on or not. Therefore, you build a temperature sensor and connect it to the heater. You build this sensor in such a way that it is capable of sending a command to the heater to go on every time the temperature falls below 68°F and to shut off when it exceeds 72°F. This way, the heater is then said to be equipped with a sensor that provides the heater with **feedback** (i.e., information on the consequences related to a specific event *fed back* to the system that initiated the event). Without proper feedback, machines (and living organisms alike) would be unable to correct for errors they commit (how could you judge your academic performance in school without feedback in the form of a report card?). Thus, feedback is necessary for successful adaptation to changes in the environment.

Therefore, the essential role that the cerebellum seems to play in the successful execution of voluntary motor commands is coordination through feedback. Improper feedback leads to inability to adjust for errors in motion in time. This becomes most obvious when intention tremor intensifies toward the completion of the movement, as the limb oscillates more vigorously closer to the target intended to be reached. However, in the last two decades, many amendments to this theory have been proposed and were actually recently accepted, including the inclusion of cognitive and linguistic coordination in the list of functions mediated by the cerebellum. The most drastic challenges to the notion that the cerebellum plays only a motor role come from experiments showing cerebellar abnormalities linked to psychiatric disorders as well, such as schizophrenia and autism. These recent findings have led to a serious evaluation of the general theory of cerebellar motor function. The debate currently focuses on the general role that the cerebellum plays in the coordination of many different brain processes. This coordination appears to rely on a principal faculty of the cerebellum—that is, the ability to correct for errors before or during execution of brain commands, which are not restricted to voluntary motor commands but may extend to the sensory, cognitive, emotive, and even linguistic domains as well.

■ **Learn more about cerebellar motor function** Search the Internet for *cerebellar function* or *cognitive and linguistic deficits*.

Current Topics in Motor Learning and Intracellular Memory

8

A prerequisite for learning is memory. One simply cannot expect to learn if one is unable to commit to memory a subject or an act by storing the information (memorization) and retrieving it later upon demand (recalling stored information; remembering). Therefore, based on the time elapsed between storing an event in memory and the moment of recollection, a memory may be classified as either short-term or long-term. For example, if you've just been reading through this whole chapter, reciting the main idea introduced in this paragraph may be a good test of your short-term memory, which surely also depends on your level of commitment to learn, to pay attention to the words, and to focus on the subject (distracted readers, even if reading out loud, would not properly recall what they have just read). It is widely accepted that memory mechanisms (at least cognitive memory) are mediated by the brain. For example, **Alzheimer's disease** is a progressive brain disorder that causes neuronal death in a specific brain region (including the hippocampus), leading to a gradual decline in cognitive memory (**dementia**) and therefore a reduction in a person's ability to engage in cognitive learning, reasoning, making judgments, social communication, and carrying out basic daily activities. As Alzheimer's disease progresses, individuals may also experience changes in personality and behavior. Dementia may also have underlying causes other than Alzheimer's disease; for example, as the result of a brain stroke. But motor and cognitive memories differ in many respects, especially with regard to underlying brain mechanisms and brain regions involved.

■ **Learn more about Alzheimer's** Search the Internet for *Alzheimer's disease* or *dementia.*

COMPLEX AND SIMPLE SPIKES REVISITED

Learning how to ride a bicycle and how to swim are good examples of long-term motor memories. It is said that one never forgets how to ride a bicycle. Interestingly, the ability to ride a bicycle is retained even in patients diagnosed with Alzheimer's disease, and so are other forms of long-term motor memories such as driving a car (although the exact destination may be confused or forgotten). This further supports the argument that the neurons underlying long-term memory in its motor versus cognitive forms may be different. In fact, patients diagnosed with cerebellar abnormalities demonstrate disturbances in their motor skills as detailed in the previous chapter (intention tremor ataxia), although it is less clear whether they also "forget" long-term motor skills. What is known, however, is that it is possible to interfere with the process of motor learning in general if the cerebellar circuitry is damaged in an experimental setting. It becomes logical, then, to propose that the cerebellum not only plays a role in the coordination of many brain processes through feedback, but also underlies the basic mechanisms for motor learning. How could this be achieved at the circuitry level of the cerebellum?

Let us build a cerebellum again, with more emphasis this time on connectivity and functional implications of the synapses. That is, let's try to look further into how the cerebellar "hard-

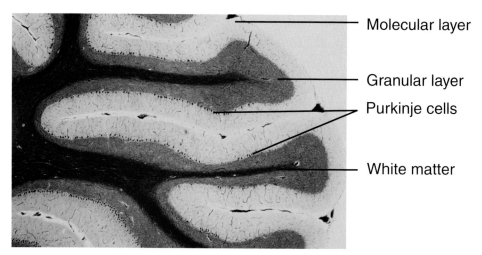

Molecular layer

Granular layer
Purkinje cells

White matter

Figure 8.1 This photograph shows a sagittal cut made through the cerebellar cortex (white matter appears darker because of the stain used). Purkinje cells are lined up like beads on a string, forming the Purkinje cell layer that serves as a boundary between the molecular layer (above) and the granular layer (below).

ware" might allow motor learning. First, we will build the outer-most part of the cerebellum, the cerebellar cortex (not to be confused with the *cerebral* cortex). As discussed earlier, Purkinje cells are the most prominent cells of the cerebellar cortex; they have the largest cell bodies and they are the only output from the cerebellar cortex, so we will start with those (Figure 8.1).

The human cerebellum has 7 million to 14 million Purkinje cells. Each receives more than 100,000 synapses on its dendritic extensions. However, the types of synapses and their location on the dendrites differ based on the source of the input. It was previously mentioned that two main types of excitatory input synapse on Purkinje cells. Here, we will further explore their source and their contribution to Purkinje cell physiology. One of these inputs originates from a nucleus in the brain stem at the level of the medulla, called the inferior olive (arguably because the outside appearance of the brain at that site resembles a pair of olives). A pair of inferior olive nuclei each contains neurons whose axons project to the contralateral cerebellum. These axons ascend toward the cerebellar cortex and synapse on the portions of the Purkinje cell dendrites closest to the cell body (**proximal dendrite**). Because they ascend all the way from the inferior olive in the brain stem to form a complex network of synapses with the Purkinje cell dendrites, they are referred to as the climbing fibers. Each climbing fiber "climbs" toward only one contralateral Purkinje cell. However, one climbing fiber forms as many as 26,000 individual synapses with the proximal dendritic extensions of one Purkinje cell, thus forming one of the most powerful neuron-to-neuron contacts in the whole nervous system. As a result of this powerful and intimate contact, each time a signal (an action potential) is transmitted to a Purkinje cell via this pathway, a distinct action potential is evoked in the Purkinje cell. In fact, this action potential is so distinct that it can be recorded only from nearby (or within, if it's an intracellular recording) a Purkinje cell in the cerebellum. Purkinje cell action potentials evoked by climbing fiber synapses are referred to as complex spikes, and complex spikes are therefore a "trademark" or an electrophysiologic signature for Purkinje cells.

The second input to Purkinje cells originates from neurons within the cerebellar cortex, below the Purkinje cell layer (the layer formed by the cell bodies of Purkinje cells lined up across the whole cerebellum). These neurons are called granule cells (presumably because they have a grainy appearance when viewed under a microscope). Granule cells are the most abundant type of cells in the cerebellar cortex. Their axons ascend toward the surface of the cerebellum beyond the Purkinje cell layer, then split, or bifurcate, to form two branches traveling in completely opposite directions along a longitudinal axis (left to right or vice versa). When axons of granule cells split at the superficial layers of the cerebellar cortex, they become known as parallel fibers, forming bundles of fibers that run in parallel.

The functional implication of this bifurcation is the ability to communicate information to more than one Purkinje cell at a time. Accordingly, an action potential ascending toward the surface of the cerebellum via a granule cell axon will evoke two action potentials at the bifurcation point, traveling in opposite directions horizontally via parallel fibers. In fact, one parallel fiber travels for a relatively long distance throughout the cerebellar cortex, and synapses on the distal dendrites of many Purkinje cells along its path (almost 200 Purkinje cells). This is how one action potential in a granule cell could hypothetically initiate multiple action potentials in Purkinje cells along the paths of its parallel fiber, forming a strip, or a **beam of excitation**. However, parallel fibers—Purkinje cell synapses, unlike those made by a climbing fiber—are less powerful. It takes activity in more than one parallel fiber for the Purkinje cell to reach its threshold of activation. Such action potentials (evoked by parallel fibers) are called simple spikes (in contrast to complex spikes) and bear more resemblance to the rest of the action potentials encountered throughout the nervous system.

Other types of neurons can also be found in the cerebellar cortex, but these are mostly inhibitory. Some of these interneurons are interjected between adjacent parallel fibers. As a result of their inhibitory synapses, they cause inhibition laterally in the surrounding Purkinje cells that they contact, forming a typical pattern encountered here and elsewhere throughout the central nervous system, called **surround inhibition**. As a strip or a beam of parallel fibers is excited ("on" beam), adjacent strips are inhibited due to the interneurons that are interjected within the superficial horizontal circuitry, forming a surrounding zone of inhibition ("off" beam). The surround inhibition phenomenon in the nervous system is thought to amplify the effect of excitation within a certain area by contrasting it with an inhibited surrounding. This could be illustrated by reading letters in a dark color font over a dark background versus a lighter background (contrast usually enhances visual acuity). It is also thought that surround inhibition helps the brain better "focus" on the information processing of signals within an active zone.

LTD IN PURKINJE CELLS

Normally, the frequency of complex spikes is approximately 1–10 Hz (or Hertz, which is the number of events per second). Simple spike frequency, however, is much higher, typically 10–60 Hz, but can often reach more than 80 Hz. These two types of activity (based on the two sources of input) can occur within the same Purkinje cell. Their pattern has been extensively studied, especially in relation to each other. For example, as a Purkinje cell fires simple spikes continuously and spontaneously, every time a complex spike occurs, a brief period of silence occurs (typically lasting 100–200 milliseconds) within the cell before it picks up simple spike firing again. However, under certain experimental conditions, simple spike frequency can be drastically decreased for a long period of time. Such a decrease in activity is referred to as **long-term depression** (LTD). As discussed earlier in this chapter, in relation to cognitive or motor memory, a distinction can also be made here between long- versus short-term depression, depending on the period between storage and retrieval of the message. Other forms of changes in neuronal activity pattern include short- and long-term potentiation (LTP, or increase in activity). It is highly speculative to say that these forms of cellular memory may well underlie the general memories we see in the behavior of living organisms. Furthermore, focus has now shifted to the study of intracellular mechanisms at the molecular level that permit these sustained memories. However, prior to exploring the biochemical pathways, it is worth discussing the experimental setting leading to LTD in Purkinje cells and its possible implication in motor learning and memory.

Suppose we insert three microelectrodes into the cerebellum. The tip of a recording electrode (RE) is placed near the Purkinje cell layer to record neuronal activity near the cell body of a Purkinje cell. Another electrode is placed toward the superficial layer of the cerebellar cortex, along the path of parallel fibers. This stimulating electrode (SE1) is used to stimulate parallel fibers along the same longitudinal line as the recording electrode. A second stimulating electrode (SE2) is placed within or near the contralateral inferior olive (the origin of the climbing fibers). This electrode is also used for the stimulation of most climbing fibers that originate from that inferior olive nucleus, in particular the one climbing fiber connected to the Purkinje cell recorded by the RE. Even before any stimulation, the *spontaneous* neuronal activity recorded through the RE will show a typical firing pattern: Up to 60 Hz of simple spikes are interrupted by brief periods of silence after every occurrence of a complex spike (which usually occurs at a frequency of 1–10 Hz). However, artificial electrical stimulation through SE1 will result in an additional simple spike activity, whereas stimulation through SE2 will evoke a complex spike because the origin of the climbing fibers within the inferior olive were stimulated. If random, these artificial stimulations are not expected to interrupt the typical firing pattern of a Purkinje cell. However, something interesting happens when parallel fibers and climbing fibers are stimulated simultaneously and repeatedly. Parallel fiber synapses became less effective, and the general activity level in the Purkinje cell is decreased drastically. The Purkinje cell enters into a profound state of LTD. This LTD period has been reported to last for more than a few days in an experimental setting, depending on the frequency and pattern of stimulation of both types of input (parallel fiber/climbing fiber) and how long the cells can be kept alive in the laboratory.

Could LTD in Purkinje cells be considered a form of cellular memory? How could this experimental finding be incorporated into the general framework of cerebellar motor learning? Some of these answers may lie in the connections that Purkinje cells make deep within the cerebellum itself, where the main target of the Purkinje cells' projections are buried, deep in the cerebellar nuclei. In fact, experimental and selective lesions of specific deep cerebellar nuclei prevent animal subjects from learning certain motor tasks.

THE CEREBELLAR CIRCUITRY SUPPORTS MOTOR LEARNING

To better understand the physiological significance of the cerebellum in terms of its functions, whether the intention is to adjust or coordinate ongoing movements and speech or to learn a new motor task, it is essential to put the above-mentioned details about Purkinje cell input (from parallel fibers and climbing fibers) and Purkinje cell output (toward deep cerebellar nuclei and brain stem nuclei) into the context of a neuronal network organization. How is input to the cerebellar cortex processed into a cerebellar output?

It was mentioned earlier that the overwhelming majority of neuronal cell bodies in the cerebellum are located within the cerebellar cortex. Because of the anticipated functions of the cerebellum, it is logical to assume that the input relayed to the cerebellum actually carries information that is mostly related to motor commands from the *cerebral* cortex (indirectly through a relay in the brain stem in the pontine nuclei). That information allows the cerebellum to process the "plan of action" before, or even during, the execution of a voluntary command, such as reaching for an object, throwing a baseball, or articulating words. Relevance to the topic of feedback, also previously discussed, comes more into light here. If the cerebellum plays a central role in the correction of error signals (generated from improper, incomplete, or faulty execution of a voluntary task), it must receive the plan of action ahead of time, or at least while the task is being carried out. It then sends "recommendations" back to the cerebral cortex, the ultimate "commander" of voluntary actions (Figure 8.2). This continuous feedback to the cerebral cortex is not direct; it is relayed through brain nuclei in the **thalamus** (the input to the cerebellum from the cerebral cortex is also indirectly relayed via pontine nuclei). However, even if the plan of action presumably matches the executed actions, information must also be received from peripheral sensory receptors. This information is necessary for the cerebellum to "compare" the status of the peripheral body parts (e.g., the limbs or the trunk) with the intended cerebral command. This feedback from the spinal cord ascends to the cerebellum in distinct tracts in the spinal cord, depending on the type of information carried. For example, a specific tract relays information about joint position and movement in space, whereas other tracts relay information about touch, pressure, or pain from the skin. These tracts form a complex network of pathways referred to as the spinocerebellar tracts. Approximately 20 such tracts have been identified. Briefly, a minority of these tracts carries information primarily related to peripheral events from muscle, joint, and skin receptors, the result being a global representation of limb parameters (e.g., reflexes and stepping or scratching patterns) rather than a muscle-by-muscle or joint-by-joint representation (as in the case of other sensory tracts "specialized" for mediating accurate sensory perceptions).

All input to the cerebellar cortex, whether it is indirect from the cerebral cortex or direct from the spinal cord, can only reach Purkinje cells through one of the two pathways: parallel or climbing fibers. The brief pause in Purkinje cell firing after a complex spike is thought to be an error signal, "alerting" the cerebellum to a faulty execution in a voluntary command. The Purkinje cell then pauses for a while and picks up firing subse-

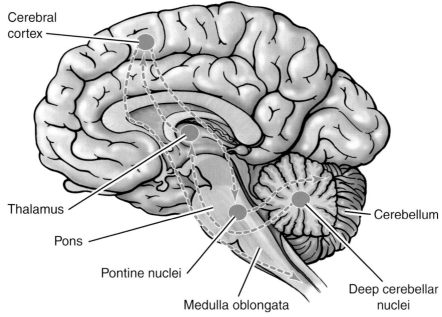

Figure 8.2 The information for voluntary command starts in the cerebral cortex (motor). It is then relayed through the pontine nuclei to the cerebellar cortex and to the cerebellum. The information then travels back to the cerebral cortex through the thalamus. This loop is thought to help coordinate and "fine-tune" a voluntary command before it ultimately arrives in the spinal cord.

quently. The pausing period depends largely on how severe the error is, which is thought to be translated as a change in the frequency and pattern of complex spikes, which could subsequently result in LTD in the Purkinje cell. According to this hypothesis, which is currently under intense experimental investigation, the inferior olive (origin of the climbing fibers) plays the role of an "error sensor" to Purkinje cells. Being the only output from the cerebellar cortex, the Purkinje cells then "reflect" these error codes on their targets that lie deep within the cerebellum, i.e., to neurons clustered in the deep cerebellar nuclei (in addition to other neurons in other nuclei in the brain stem called vestibular nuclei, which are involved in eye movements and balance). These are a group of three nuclei, each mediating different functions depending on their output connections with the brain and the brain stem. In any case, the deep cerebellar nuclei are a major output from the cerebellum. That is, most cerebellar functions are mediated through the deep cerebellar nuclei (exceptions to this rule are again the vestibular nuclei, which receive direct synapses from Purkinje cells and coordinate eye movements and balance).

However, the picture is not yet complete. Purkinje cells release an inhibitory neurotransmitter from their axonal terminals. Therefore, every time a Purkinje cell is "activated" or fires an action potential down its axon, a target neuron in a deep cerebellar nucleus (or in the vestibular nucleus) is inhibited, and the general cerebellar output is consequently silenced, correcting for the error. Accordingly, it is as though neuronal activity in the Purkinje cells in the cerebellar cortex is reflected in a mirror-image fashion in the deep cerebellar and vestibular nuclei. This was eloquently described by British physiologist Charles Sherrington in 1904: "The cerebellar cortex is an enchanted loom where millions of flashing shuttles weave a dissolving pattern, always a meaningful pattern though never an abiding one; a shifting harmony of subpatterns." Sir Charles Sherrington's research, spanning more than 50 years, laid the foundations for modern neurophysiology, proposing that the most important function of the nervous system in higher animals is the coordination of the various parts of the organism. He is best known for his studies on spinal reflexes, making equally great contributions to the physiology of perception and behavior in general. He was the first to adequately study the synapse and actually originated the term. In 1932, he shared the Nobel Prize in Physiology or Medicine with Edgar Douglas Adrian.

Many years later in 1967, Sir John Eccles, who studied under Sir Charles Sherrington, reiterated:

The negative image of the integrated output from the cerebellar cortex is, as it were, formed by a process analogous to the sculpturing in stone. Spatio-temporal form is achieved from moment to moment by the impression of a patterned inhibition upon the "shapeless" background discharges of the subcerebellar neurons, just as an infinitely more enduring form is achieved in sculpture by a highly selective chiseling away from the initial amorphous block of stone.

Although Sir John Eccles was born in Australia in 1903, he entered Oxford University in the United Kingdom as an undergraduate in order to study under Sir Charles Sherrington. However, he owes much for his early training to his father and his mother. He was awarded a Doctorate in Philosophy (Ph.D.) degree in 1929 for his thesis on excitation and inhibition. Sir Eccles shared the 1963 Nobel Prize in Physiology or Medicine with two other prominent scientists, Alan Hodgkin and Andrew Huxley.

INSIDE NEURONS

So far, neurons have only been presented from the points of view of structural anatomy and general physiology. However, neurons, just like other living cells in the body, have elaborate intracellular molecular machinery within their cell bodies, as well as within the intracellular space of their dendritic and axonal extensions. Their basic molecular design includes a nucleus and other intracellular organelles and compartments. In general, the term *intracellular* refers to the space inside the membrane surrounding the entire cell (extracellular membrane) and sealing it from the outside environment. At certain spots or intervals, this membrane seal is not completely impermeable to certain molecules, to allow for the selective exchange of gases, nutrients, and by-products. This exchange may happen at specific zones through "pores" in the membrane, and these pores are typically referred to as "channels." Membrane channels in the neuron allow passage of charged molecules (**ions**) across the membrane. For example, when a sodium molecule (Na) loses an electron, it becomes a positively charged ion (Na^+). Specific membrane channels in neurons that allow only Na^+ to cross (usually from the extracellular to the intracellular space) are therefore called sodium channels. These channels are distributed close to the cell body and along the axon (if the axon is myelinated), they cluster mainly at the **nodes of Ranvier.**

Ions are usually segregated across both sides of the membrane and maintain a certain equilibrium. This separation of charges creates a **voltage difference** across the membrane (voltage difference is created by the uneven distribution of electric charges across a selectively permeable membrane or barrier). In addition to the separation of charges across the membrane, an additional factor determines the voltage difference across the membrane: the **chemical gradient** (which also develops due to the difference in molecular concentrations across a selectively permeable membrane or barrier). But when ions of similar charges cross the membrane, they create a current that causes a change in the membrane voltage.

Voltage difference is defined as the difference in potential between two points (across a neuronal membrane) due to accumulation of charged particles (ions) on either side of an impermeable barrier (the neuronal membrane). When the barrier allows ions carrying these electric charges to cross (by opening pores across the membrane, otherwise called membrane ion channels), the movement of these ions generates current that spreads along the axon, creating an "action potential."

At any rate, these changes in voltage across the neuronal membrane underlie the initiation and propagation of action potentials in neurons. Sodium channels mediate these voltage changes (along with other membrane channels) by contributing in part to the passage of current across the membrane due to unidirectional crossing of Na^+ to the intracellular space. In fact, specific sodium channel blockers could be used to inactivate neurons that express these channels. A good example to illustrate this clinically is when the dentist injects a local anesthetic before pulling a tooth. The anesthetic does not normally spread to other parts of your body or even to your whole mouth, but rather acts close to the injection site. The anesthetic acts rapidly (within 1–10 minutes), with reversible short-acting effects and virtually no side effects. The underlying mechanism at a cellular level is nothing but an inactivation of sodium channels expressed extracellularly on neuronal membranes within the injection site, thus blocking conduction of action potentials along axons or simply preventing their initiation. Numbness and loss of pain or touch sensations (anesthesia) result simply from inactivation of these neurons that carry sensory signals. On the other hand, if loss of motor function also occurs—for example, paralysis in the tongue—then the anesthetic might have spread to other neurons that carry motor signals as well (in other words, the tip of the needle may have caused direct damage to a branch of a motor nerve, and this is less rapidly reversible, depending on the extent of the damage).

It becomes obvious, then, that to better understand neuronal physiology, a closer look at intracellular events is necessary, especially when discussing cellular memory in the form of LTD. Although the inability to demonstrate similar forms of LTD in awake, behaving animals remains a serious caveat in linking molecular forms of memory to behavioral forms, discussion of molecular pathways is warranted because they may be also encountered elsewhere in the nervous system and in many non-neuronal cells as well.

LEARNING AT THE MOLECULAR LEVEL

LTD in Purkinje cells is induced by coincidence of an event that follows impulses in parallel fibers (simple spike) and another that follows those in climbing fibers (complex spike). Various forms of LTD can be induced depending on the pattern of these synchronous activations. Even though these findings have not been shown to occur naturally (without artificial electric stimulation), and even though LTD in Purkinje cells is one of the molecular bases of motor learning and memory in the cerebellum, similar molecular reactions also occur elsewhere in the nervous system.

Although both parallel fibers and a climbing fiber converge on one Purkinje cell, their synapses occupy different spaces on the Purkinje cell dendrites. Parallel fiber synapses occupy the more distal parts, whereas climbing fibers occupy those that are more proximal. Furthermore, during LTD each may release a different neurotransmitter and thus impact the postsynaptic membrane (on the Purkinje cell) differently. One neurotransmitter that is released at both types of synapses is **glutamate**. Glutamate is present in most tissues. It is actually one of the most common amino acids found in nature as the main component of many proteins. Glutamate is also produced in the body and plays an essential role in human metabolism. However, it also plays a primary role in the central nervous system as an excitatory neurotransmitter. Its receptors on the postsynaptic membrane can be of different types. Some glutamate receptors, for example, are similar to channels that, once activated by glutamate, allow ions to cross the extracellular membrane (ionic or ionotropic receptors). Others are linked to other proteins on their intracellular domain that, once activated by glutamate on the extracellular surface, trigger a chain of molecular reactions intracellularly. As a result, ionic receptors tend to have faster intracellular consequences than other types of receptors: Channels open and ions flow in almost instantaneously, whereas for other types of receptors, there is a longer delay before the consequences of receptor activation can be observed (in addition to other differences well beyond the scope of this discussion).

Upon activation, one type of glutamate ion channel becomes permeable to ions carrying positive charges including Na^+, which then causes a transient increase in intracellular Na^+ concentration and a rapid change in membrane voltage. The postsynaptic membrane of the Purkinje cell expresses another type of ion channels: calcium channels. These calcium channels act as "voltage sensors." When glutamate receptors are activated and allow Na^+ to cross the membrane, the change in membrane voltage thus created by the transient sodium current activates calcium channels. When calcium channels are activated, they in turn similarly allow calcium ions (Ca^{++}) inside the cell (calcium ions are formed when the calcium molecule loses two electrons; hence, the double positive charge and symbol). However, these three sequential steps discussed so far (release of glutamate, activation of glutamate receptors, activation of calcium channels) are only a few of the many reactions that occur normally at these synapses. What makes this relevant to LTD is the way in which these steps occur (e.g., the extent of glutamate release directly related to the intensity of stimulation in the presynaptic terminal of parallel or climbing fibers, and the subsequent amount of calcium triggered intracellularly). More than 30 intracellular molecules and many neurotransmitters have been identified in the molecular pathways leading to LTD. In the end, one major question remains largely unanswered: How do these molecular reactions retain the memory of a physiological event in the synapse? How does the Purkinje cell "remember" past experience after the neurotransmitter has been released?

NEUROSCIENCE: FROM BEHAVIOR, TO MOLECULES, TO MIND

Apparently, the cell is able to integrate this molecular information, at least in part, thanks to the intracellular signaling of calcium. Just as some types of calcium channels act as membrane voltage sensors, other intracellular proteins are very sensitive to fluctuations in calcium concentrations. These calcium-sensing proteins then transfer this information inside the cell nucleus where the genetic material is stored in the form of DNA (deoxyribonucleic acid). In turn, DNA controls the rate and type of protein "manufacturing." Therefore, there is a link between a physiological event occurring at the synapse (that is, a reflection of the intensity of neuronal activity) and the translation of this event into molecular signals that impact protein expression and, ultimately, cellular function at many levels.

In broader terms, we have outlined here a hypothetical mechanism to a behavioral observation, such as long-term motor memory invoked when riding a bicycle or playing a favorite piano piece, and pinned it down gradually to action potentials, neurotransmitters, and intracellular signaling molecules, leading to cellular changes at a genetic level. That, even if still highly speculative, is a fascinating aspect of neuroscience: the ability to link behavior to biology that may one day lead to unlocking the mysteries of the human mind.

Glossary

Action potential An all-or-none wave of electrical excitation that triggers the release of neurotransmitter from nerve terminals.

All-or-none response An event (i.e., action potential) that, when it occurs, has invariable characteristics.

Alzheimer's disease Neurodegenerative disorder causing progressive brain damage in a specific brain region (including the hippocampus), leading to dementia and gradual decline in cognitive memory.

Ammeter Device for measuring current (measured in amperes).

Antagonistic muscles A group of muscles acting to pull a limb in opposite directions along a joint (usually receiving coordinated neuronal innervation).

Artery Blood vessel carrying blood away from the heart (therefore relatively more oxygenated than blood in the veins).

Ataxia Disordered voluntary motor performance.

Autonomic Autonomous, in reference to the unconscious autonomic nervous system.

Axonal regeneration When axons are damaged following injury, they attempt to regrow if the cell body is spared. In the peripheral nervous system, axons may regenerate, forming new synapses. However, axons in the central nervous system are less likely to regenerate.

Beam of excitation A strip of action potentials spreading across a bundle of axons.

Binary code Two symbols used to transmit a message.

Brain stem Structures at the base of the brain separating the brain from the spinal cord. They include the medulla, the pons, and the midbrain (on a caudo-rostral axis), but not the cerebellum.

Central nervous system The brain and the spinal cord.

Central sulcus Prominent trough in the folding of the brain, separating between two gyri (peaks) on each side called the precentral (motor) and the postcentral (sensory) gyrus.

Cerebellar peduncles The only bundles of fibers resembling bridges and connecting the cerebellum to the rest of the nervous system. They include three peduncles on both sides of the midline (six total).

Cerebral motor cortex Brain region containing neurons mediating voluntary motor commands; damage to these neurons leads to paralysis of the corresponding body part mainly contralateral to the brain lesion.

Chemical gradient Difference in concentration of molecules across a membrane, creating a chemical force acting to restore equilibrium.

Climbing fiber Axonal extension of a neuron in the inferior olive nucleus in the brain stem and ascending to synapse on a Purkinje cell.

Complex spike Action potential with a primary peak and secondary smaller peaks evoked in a Purkinje cell following excitation by a climbing fiber.

Cranium Skull.

Dementia Memory loss.

Dendritic tree Extensions from dendrites in a neuron. Typically where neuron receives chemical signal through synapse.

Dorsal horns Gray matter in dorsal spinal cord containing mainly neuronal cell bodies receiving sensory input from the periphery.

Effector organs End target of a neuron, for example, a gland (autonomic neuron) or a skeletal muscle (motor neuron).

Electrophysiology Using electrical equipment (for e.g., amplifier, electrode, oscilloscope) to study and record neuronal activity.

Encode Receive a message or a code.

Feedback A reaction or response relayed back to the source of the action.

Folia Foldings in the surface of the cerebellum.

Frequency Number of events in time, usually reported in Hertz (Hz, number of events per second).

Functional magnetic resonance imaging (fMRI) An imaging device that shows changes in blood flow to particular areas of the brain.

Glia Supporting cells of the central nervous system providing a line of defense against invading pathogens (bacteria, viruses, etc.) and maintaining favorable cellular environment or homeostasis (recycling neurotransmitters, regulating pH, etc.), including myelin formation.

Glutamate One of the most common amino acids found in nature as the main component of many proteins. Glutamate plays a primary role in the central nervous system as an excitatory neurotransmitter.

Gyrus Peak in the folding of the brain between two sulci.

Ions Molecules carrying one or more positive or negative charge, for example sodium (Na^+), potassium (K^+), calcium (Ca^{++}) or chloride (Cl^-) ions.

Ipsilateral On the same side.

Long-term depression A long-lasting inhibition.

Multiple sclerosis Disease affecting the central nervous system (brain, brain stem, and spinal cord) typically at an early age during adulthood with recurrent episodes of recovery but gradual worsening in severity. Mostly fatal, characterized by motor deficits and sensory impairment in different body parts, due to multiple lesions referred to as 'plaques' spreading throughout the central nervous system (hence 'multiple sclerosis').

Myelin A fatty substance that enhances neuronal conduction.

Neuron Underlying cell component of the tissue of the nervous system.

Neurotransmitter Molecule synthesized and secreted by a neuron with specific receptors and mechanisms for removal or recycling upon release.

Nodes of Ranvier Normal interruption in myelin along a myelinated axon.

Nuclei (singular is *nucleus*) Groups of neuronal cell bodies, usually serving comparable functions.

Oscilloscope An instrument that graphically displays changes in electrical voltage and current.

Parallel fibers Axonal extensions of neurons originating from all bodies in the cerebellar cortex. These axons travel in parallel on the surface of the cerebellum contacting Purkinje cells.

Paralysis Loss of motor control.

Parkinson's disease A brain disorder occurring when certain neurons in a part of the brain called the substantia nigra die or become impaired. Normally, these cells produce a vital chemical known as dopamine that allows smooth, coordinated function of the body's muscles and movement. When approximately 80% of the dopaminergic neurons degenerate, symptoms of Parkinson disease start to appear.

Poliomyelitis Viral disease that is uncommon these days, largely because of vaccines administered at an early age. In the early to mid-20th century, before the vaccine was developed, children suffered fever and other typical signs of the disease and eventually experienced severe motor deficits or even complete paralysis.

Pontine nuclei Nuclei, or groups of cell bodies, located within the pons.

Postmortem Examination after death or autopsy.

Proximal dendrite Part of the dendrite closer to the cell body.

Purkinje cells Nerve cells that form a layer in the cerebellum.

Receptors Proteins expressed usually on the postsynaptic membrane of a neuron. Primary site of action of a neurotransmitter.

Regenerate Repair oneself; regrow.

Reticular formation Aggregation of cell bodies interspersed and not forming a distinct nucleus.

Retina Membrane in the eye impacted by light and containing light-sensitive neurons (photoreceptors).

Simple spike Action potential evoked by synapse of parallel fiber on a Purkinje cell.

Skeletal muscles Muscles of the body responding to voluntary motor commands (smooth muscles of the viscera are under autonomic control).

Somatosensory Sensation from the soma (the body), in contrast to visceral (from viscera).

Spinal cord injury Direct damage to spinal cord tissue typically caused by violent accidents, or indirect, following the injury due to interrupted blood supply and severe inflammation.

Stroke Sudden interruption of blood supply to neurons in the brain, typically due to blood clot occluding a blood vessel.

Sulcus Trough in the folding of the brain between two gyri.

Surround inhibition A region of inhibition surrounding another activated region.

Synapse Neurons come into close contacts with each other to transmit signals, through a synapse, mainly at the end of an axon of one neuron and the dendrite of another (referred to as axo-dendritic synapse, or less commonly between an axon and another referred to as axo-axonic).

Synaptic zones Places where neurons make contact with one another.

Thalamus Almond-shaped structure approximately in the center of the brain, a necessary relay for ascending information (mainly somatosensory).

Tracts Group of axons traveling in parallel and serving similar function(s).

Ventral horns Gray matter in ventral spinal cord containing mainly neuronal cell bodies sending output to target peripheral organs (motor neurons if the target is a muscle, or autonomic neurons if the target is a visceral organ).

Vermis Worm-like structure midline in the cerebellum between both cerebellar hemispheres.

Visual cortex Gray matter in the brain containing cell bodies of neurons receiving visual input.

Voltage difference Separation of charges (ions or electrons) across a barrier medium (or a membrane in the case of neurons) creating an electric force acting to restore equilibrium.

Wallenberg syndrome Neurological disorder generally caused by a blockage in an artery supplying part of the medulla and the inferior aspect of the cerebellum. Patients typically have swallowing difficulties and hoarseness, loss of taste, and some paralysis of facial muscles. Symptoms may also include dizziness and loss of pain or temperature sensitivity.

Bibliography

Johnston, D., and S. M. Wu. *Foundations of Cellular Neurophysiology.* 2nd ed., Cambridge, MA: MIT Press, 1995.

Kandel, E. R., J. H. Schwartz, and T. M. Jessell. *Principles of Neural Science*, 4th ed. New York: McGraw Hill, 2000.

Nolte, J., and J. B. Angevine. *The Human Brain.* St. Louis: Mosby, 1995.

Rhawn, J. *Neuropsychiatry, Neuropsychology, Clinical Neuroscience.* New York: Academic Press, 2000.

Further Reading

Butler, A. B., and Hodos W. *Comparative Vertebrate Neuroanatomy: Evolution and Adaptation*. New York: Wiley-Liss, Inc., 1996.

Hendelman, W. J. *Atlas of Functional Neuroanatomy*, Boca Raton, FL: CRC Press, 2000.

Johnstone, A. *"So, What's Really Going on in Those Young Heads? They May Look Like Adults, but Don't be Fooled. New Studies Show the Brain is Such a Cauldron of Activity that our Teenagers are Far from Mature."* The Herald. *(UK)*: Jan 10, 2004. p. 4.

Marshall, L. H., and Magoun H. W. *Discoveries in the Human Brain: Neuroscience Prehistory, Brain Structure, and Function*. Totowa, N.J.: Humana Press, 1998.

Paxinos, G., and Mai J. K. (Editors). *The Human Nervous System*, 2nd ed. San Diego: Academic Press, 2003.

Squire, L. R., et al. *Fundamental Neuroscience*, 2nd ed. San Diego: Academic Press, 2003.

Websites

Alzheimer's Association
http://www.alz.org/

BioResearch
http://bioresearch.ac.uk/

Brain Explorer
http://www.brainexplorer.org/

Cleveland Clinic (on inflammation)
http://www.clevelandclinic.org/health/health-info/docs/0200/
0217.asp?index=4857

Latin and Greek words
http://www.answers.com/topic/list-of-latin-and-greek-words-
commonly-used-in-systematic-names

Latin name neuroanatomy
http://www.neuropat.dote.hu/anastru/elist2.htm

National Multiple Sclerosis Society
http://www.nationalmssociety.org/What%20is%20MS.asp

National Parkinson Foundation
http://www.parkinson.org/site/pp.asp?c=9dJFJLPwB&b=71117

Neuroscience for Kids
http://faculty.washington.edu/chudler/neurok.html

The nervous system (emc)
http://www.emc.maricopa.edu/faculty/farabee/BIOBK/BioBook
NERV.html

New Horizons for Learning
http://www.newhorizons.org/neuro/leiner.htm

Nucleus Medical Arts
http://catalog.nucleusinc.com/catalogindex.php

Virtual Hospital
http://www.vh.org/welcome/aboutus/index.html

WhoNamedIt
http://www.whonamedit.com/

WHO on Poliomyelitis
http://www.who.int/topics/poliomyelitis/en/

Index

About the Author

Carl Y. Saab is an active neuroscience researcher. He graduated in 1997 from the American University of Beirut (AUB) in Lebanon with an M.S. in neuroscience. At AUB, he learned the principles of pain research, devoting his time to working closely with Dr. Nayef Saadé in the Department of Human Morphology. He then traveled to the University of Texas Medical Branch in Galveston, Texas, to pursue graduate studies under the guidance of Dr. William D. Willis, an internationally renowned neuroscientist in the field of pain research. He obtained his Ph.D. in 2001. His decision to join the Department of Neurology at Yale University as a postdoctoral fellow in 2001 soon put him on an exciting path toward better understanding the basic mechanisms of neuronal degeneration and reversal. He wrote this book partly while at Yale in the laboratory of Dr. Stephen Waxman, chairman of Yale's Neurology Department. The final stages of the book were written while at Brown University and Rhode Island Hospital as assistant professor of research in the Department of Surgery. Saab enjoys science and solving riddles, philosophy, and painting. Occasionally, you might also see him cruising on his road-racing bike at 20 miles per hour. Saab also enjoys working as a disc jockey in dance clubs.

Picture Credits

page:

2: ©Peter Lamb
3: ©Peter Lamb
6: Greg Gambino / 20.64 Design
7: Greg Gambino / 20.64 Design
8: National Library of Medicine
12: Mark Harmel / Photo
    Researchers, Inc.
17: Geoff Tompkinson / Science
    Photo Library
18: Philip Ashley Associates

19: Medical Art Service / Photo
    Researchers, Inc
20: Medical Art Service / Photo
    Researchers, Inc
23: Philip Ashley Associates
30: Kenneth Eward / BioGrafx /
    Photo Researchers, Inc.
56: Astrid & Hanns-Frieder Michler
    / Science Photo Library
63: Philip Ashley Associates